EXPERT SYSTEMS TECHNOLOGY

INFORMATION TECHNOLOGY AND SYSTEMS SERIES

Editor in Chief:
Professor F H George

Other titles in the series in preparation:

ON SYSTEMS
A. Clementson

THEORY AND PRACTICE OF COGNITIVE SCIENCE
Jagodzinski

THEORY AND KNOWLEDGE OF ARTIFICIAL INTELIGENCE
Johnson & Hartley

A PERSPECTIVE ON INTELLIGENT SYSTEMS
Kohout

KNOWLEDGE REPRESENTATION IN MEDICINE AND CLINICAL BEHAVIOURAL SCIENCE
Kohout & Bandler

INTRODUCTION TO LOGIC FOR SYSTEMS MODELLING
V. Pinkava

EXPERT SYSTEMS TECHNOLOGY

A GUIDE

L. JOHNSON

E.T. KERAVNOU

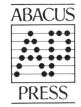

ABACUS

PRESS

Tunbridge Wells ■ London

■ Dover, New Hampshire, USA ■

First Published in 1985 by

ABACUS PRESS

51 Washington Street, Dover, New Hampshire 03820, USA

Abacus House, Speldhurst Road, Tunbridge Wells, Kent TN4 OHU, UK

NOTE: Abacus Press US office now at:
P.B.S, PO BOX 643, Cambridge, MA 02139, USA.

 British Library Cataloguing in Publication Data

Johnson, L.
 Expert systems technology : a guide.——
 (Information technology & systems series)
 1. Expert systems (Computer science)
 I. Title   II. Keravnou, E.T.   III. Series
 001.64   QA76.9.E96

 ISBN 0-85626-446-6

Printed in Great Britain by Nene Litho and bound by Woolnough Bookbinding
both of Wellingborough, Northants.

CONTENTS

for Prof. F. H. George
Guide, Counsellor and Friend

PREFACE

There is considerable interest in Advanced Information Technology and this book may be a useful guide to all those involved, be they computer scientist, electrical engineer, cognitive scientist or those orientated towards applications. Any of these may feel in need of more deep descriptions of expert systems than are usually available but yet do not need these descriptions entangled with implementation details and the idiosyncratic terminology of the systems designers. These same needs are felt by the neophyte knowledge engineer and hence it is hoped that the committed will find this book a challenging manual. Finally, we have found that our own draft documents have proved useful as a reference work and the book may also serve this purpose.

In this book we describe several expert systems at a conceptual level. We bring out their novel architectural features which gave rise to our selecting them for inclusion in the book. Each chapter has an overview, a description of the system's static structure and a description of its dynamics. We have not used the idiom of the original system constructors but have imposed as much commonality as possible. In this way we hope that the concepts employed, either implicitly or explicitly, by them are more readily discerned and comparisons are easier to make.

We would like to thank the many researchers and others involved in the design of the expert systems, particularly those connected to the systems selected for inclusion in this book, for making the field one of interest and enduring importance. It is worth the effort in getting to know the fruits of their labour. We would like to express our gratitude to the members of the Man-Computer Studies Group (here at Brunel University) who encouraged us to believe that the style and content might be worth wider exposure than our own members.

Our thanks are due to Shirley Hatch who entered the draft copy into the SUPERBRAIN QD under the SPELLBINDER word processing package. We took up the task of editing this raw material into the form seen here. Neither Shirley nor SPELLBINDER are responsible for the errors and omissions that undoubtedly remain.

April 1984
Brunel University, London.

L J & E T K

Chapter 1
INTRODUCTION

The purpose of this introductory chapter is to orientate the reader and summarize the concepts we use throughout the book. Its purpose is not to introduce these concepts to those innocent of them. The relevant concepts fall under three of the subsections of the introduction. Section 1.1 is a discussion of interactive expert systems. Section 1.2 is a sketch of the main currently available knowledge representation schemes. Section 1.3 is an overview of the inferences employed in the discharge of the tasks typically undertaken by the systems which we have selected for discussion in subsequent chapters. The final section of the introduction previews these systems.

1.1 FOCUS OF THE BOOK

A **Knowledge-Based System** is a system which manipulates "knowledge" in order to perform a task or tasks. The knowledge in a knowledge-base, is highly structured symbolic data which represents a model of the relationships between data elements and the uses to be made of them. The performance of a knowledge-based system depends both on the quality of its factual knowledge (structure, completeness, validity, consistency, etc.) and the ways in which this knowledge is applied. (See Johnson and Hartley 1981; Addis and Johnson 1983.)

The field of **Expert Systems** is a subgroup of knowledge-based systems. A prerequisite for applying the expert systems technology to some knowledge domain is the existence of human experts for that domain. The field investigates methods and techniques for constructing Human-Computer Systems encorporating the domain specific knowledge. In a manner of speaking expert systems can thus be said to grasp fundamental domain principles (and weaker general methods), to solve complex problems and to interact intelligibly with the user (see Johnson and Keravnou 1983). And in this way, using the same manner of speaking, an expert system can be said to interpret, diagnose, predict, instruct, monitor, analyse, consult, plan or design.

Expert Consultant Systems

Currently most expert systems engage in a dialogue with the user; the computer acting as a "consultant". The computer system

suggests options on the basis of its knowledge and the symbolic
data supplied by the user. A dialogue is usually terminated when
a decision or a recommendation is reached. In this book we
consider only **Expert Consultant Systems** which designate this large
subgroup of expert systems. The term, "consultant", makes
explicit the most important and distinguishing feature of these
systems; i.e. the fact that they are **interactive** computer systems.
Figure 1.1 gives the taxonomy of Computer Systems from the
perspective of the book.

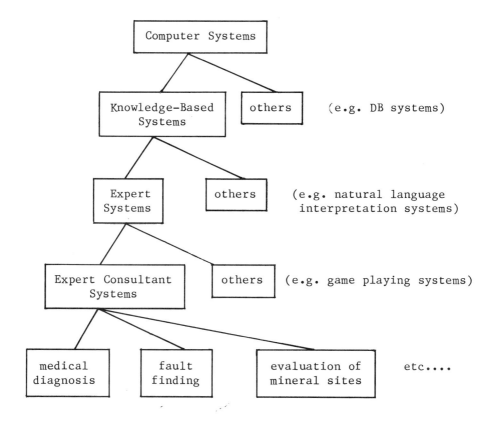

Fig. 1.1: Taxonomy of Computer Systems
from the perspective of Knowledge-Based Systems.

   The vital process of a consultant system is the process of
moving from known items of information (or "seen" concepts) to
unknown information (or "unseen" concepts). The user of an expert
consultant system has "observed" some particular state of affairs,
within the domain of the system's expertise and submits these
observations to the system. Examples of such states of affairs

are: a sick person, a faulty machine, an earth region (for
evaluation of mineral sites), a malfunctioning business environment. Based on the observations the system makes inferences and
suggests new routes of investigation which will yield high grade
information. An interaction continues until the system finds the
most likely explanation of the observations. Once a likely
explanation is reached the system may go on to compile
recommendations.

Modes of Interaction

There are two basic forms of interaction: 1) user initiated
2) computer initiated.

In the **user initiated** mode of interaction the system is
restricted to respond to user requests only. The accuracy of any
conclusion or recommendation reached by the system is constrained
by the amount of input provided by the user. The decision to
reach more accurate conclusions or recommendations lies entirely
with the user. If the user is not satisfied with the current
system output then he/she can formulate his/her next request based
on a careful examination of the output.

In the **computer initiated** mode of interaction the user is
restricted to respond to system requests only. The system starts
by a set task (usually this will be a very general task like
"Compile the best therapy regime for this patient") and in the
light of this, the system requests input that will enable it to
accomplish the task. The process is not determinate for the
system makes inferences on input and requests further input in the
light of these inferences.

In the context of expert consultant systems the desirable
mode of interaction is a **mixed-initiative** one, whereby the
initiative switches from the one basic mode to the other (both the
user and the computer prompt each other).

1.2 KNOWLEDGE REPRESENTATION SCHEMES

Knowledge can be represented in schemes that lend themselves
to implementation on a computer. The predominant schemes are:
Predicate Calculus, Associative Networks, Frames and Rules. This
section gives a very brief description of these schemes.

On the whole schemes are Janus faced; the one face looking
towards human understanding the other looking towards computer
implementation. Hence we will distinguish between Knowledge

Representation Schemes and Knowledge Representation Languages --
schemes look towards human understanding and the methodologies for
knowledge elicitation etc., **languages** are implementations of
schemes and hence questions of control and efficiencey arise.
One finds that certain languages have been directly influenced by
a scheme. A point that we wish to make is that Predicate Calculus
can be thought of as both a scheme and as the basis for a
language. When conceived as a language Predicate Calculus may be
used, as indeed any language can, to implement any of the schemes.
The argument for treating Logic Programming as a basis for work in
expert systems, is one to do with having a high level language
with a clear semantics. One can accept, or reject, these arguments independently of accepting, or rejecting, that Predicate
Calculus is the only scheme in which to consider knowledge
representation.

To aid us to make our description of the schemes we use a
very trivial body of factual knowledge which we represent in them.

"A person suffering from a disease belonging to the category of
diseases, A, exhibits symptoms X and Y. B, C, and D are such
diseases. A person suffering from B in addition to symptoms X
and Y, exhibits symptom Z as well".

The conceptual structure of the above knowledge is very
simple: we have a **taxonomy** of diseases and each node of this
taxonomy is **empirically associated** with some symptoms; a category
of diseases is associated with those symptoms shared by all its
specialisations (in other words a symptom associated with some
element of the taxonomy is inherited by all its descendant
elements (property **inheritance**)).

1.2.1 Predicate Calculus

The (first order) Predicate Calculus can be used as a basis
of a knowledge representation scheme (see Nilsson (1980), chps 4
and 5). Below we express our body of knowledge in this idiom.

1.   For all diseases x and y,
          if x is the category of y, then
               for all symptoms s,
                    if s is associated with x, then
                         s is associated with y.
becomes

1. $\forall x$ $\forall y((\text{DISEASE}(x) \wedge \text{DISEASE}(y) \wedge \text{CATEGORY}(x,y))$ $\Rightarrow$
          $(\quad \forall s((\text{SYMPTOM}(s) \wedge \text{ASSOCIATED-WITH}(s,x))$ $\Rightarrow$
               $\text{ASSOCIATED-WITH}(s,y))))$

2. For all persons p and diseases x,
       if p suffers from x, then
           for all symptoms s,
               if s is associated with x, then
                   p exhibits symptom s.

becomes

2. $\forall p \ \forall x((\text{PERSON}(p) \wedge \text{DISEASE}(x) \wedge \text{SUFFERS-FROM}(p,x)) \Rightarrow$
       $(\ \forall s((\text{SYMPTOM}(s) \wedge \text{ASSOCIATED-WITH}(s,x)) \Rightarrow$
           $\text{EXHIBITS}(p,s))))$

3., 4., 5., 6.	--- is a disease; 7,8,9 --- is a symptom;
10., 11., 12.	--- is the immediate category of --- ;
13., 14., 15.	--- is associated with --- ;

become

3. DISEASE(A)   4. DISEASE(B)   5. DISEASE(C)
6. DISEASE(D)   7. SYMPTOM(X)   8. SYMPTOM(Y)
9. SYMPTOM(Z)   10. CATEGORY(A,B)   11. CATEGORY(A,C)
12. CATEGORY(A,D)   13. ASSOCIATED-WITH(X,A)
14. ASSOCIATED-WITH(Y,A)   15. ASSOCIATED-WITH(Z,B)

These sentences, in the idiom, can be said to represent the knowledge. The specific taxonomy of diseases and the knowledge concerning the specific symptoms associated with these diseases are represented in sentences 3-15. In sentence 1, there is the explicit expression of "property inheritance" and this facilitates the representation of empirical associations between symptoms and diseases. We only need to explicitly express the associations that are particular to each disease -- the rest can be deduced.

Within this scheme, knowledge is only represented as an unstructured sequence of "independent" sentences. For example, there is no specification that sentences 3-6 and 10-12 together represent "a taxonomy of diseases". However, this grouping reflects the application of the sentences in a problem solving context and there is a need to group them explicitly to form higher level knowledge units (see Brachman **et al** 1983).

1.2.2 Associative Networks

An **associative network** is a collection of nodes and arcs (see Findler 1979). Nodes represent terms (names of physical entities, situations, places, processes, events, n-ary relationships (n >= 2), etc.) and arcs represent binary relationships or arguments of n-ary relationships (n >= 2). Figure 1.2 gives the associative network representation of our body of knowledge.

KEY:

    s: subset-of;   ds: disjoint-subset-of;
    de: distinct-element-of;   aw: associated-with.

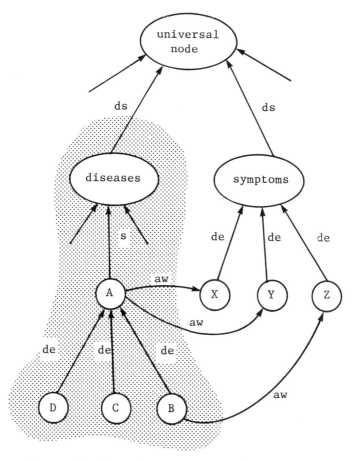

The associative networks scheme captures the given conceptual knowledge structure adequately; the taxonomy of diseases is represented through the "s", "ds" and "de" arcs and the empirical associations between symptoms and diseases are represented through the "aw" arcs.

Fig. 1.2: Illustrating the scheme of associative networks.

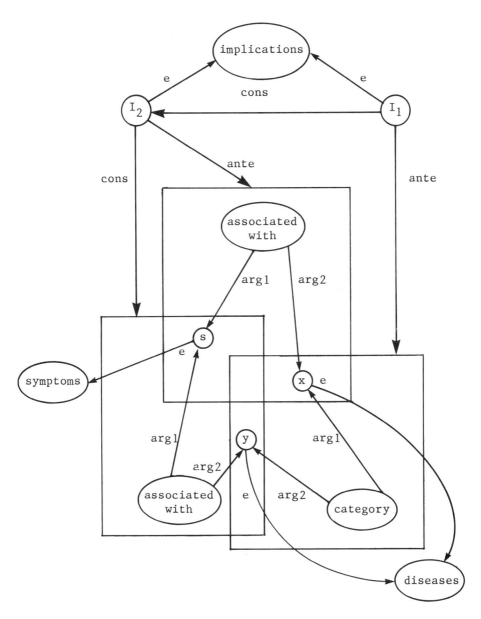

Partitioned associative network representation of the sentence
"Any disease inherits the symptoms associated with its category".
I_2 is an implication node with arcs connecting it to antecedent
and consequent. Similarly I_1 is another implication node whose
consequent is the implication represented by I_2.

Fig. 1.3

The network concept is extended by **partitioning** groups of nodes and arcs into a kind of super node (Hendrix, 1979). Figure 1.3 gives a partitioned associative network. This illustrates the point that the boxed areas represent units which are the supernodes of a super binary predicate.

1.2.3 Frames

A **frame** (Minsky 1975) is a structure consisting of a network of nodes and relations, used for representing a situation or topic stereotype. Attached to the frame is information about **how to use** the frame, what to expect to happen, and what other frames it might be appropriate to move to in certain circumstances. In short the scheme permits the co-existence of the factual knowledge and the reasoning knowledge that manipulates it. Some aspects of the frame are fixed; these are **slots** that are initially filled with **"default"** assignments containing information which holds unless new information displaces them. (See figure 1.4.)

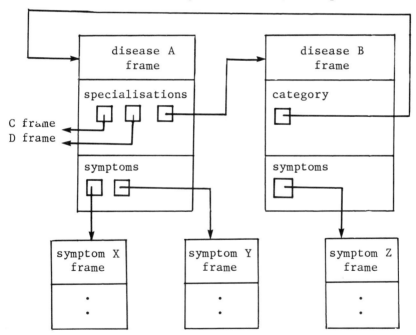

Frames also adequately capture the given structure; the taxonomy of diseases is represented through the "specialisation" and "category" slots and the empirical associations between symptoms and diseases are represented through the "symptom" slots.

Fig. 1.4

1.2.4 Production Rules

Consider the following relationships in the given body of knowledge:

patient suffers from disease A $\Rightarrow$
patient exhibits symptoms X and Y

patient suffers from disease B $\Rightarrow$
patient exhibits symptoms X, Y and Z

The knowledge is represented in rule form as follows:

R_1: **If**   the patient exhibits symptom X **and**
         the patient exhibits symptom Y
     then   the patient is <u>likely</u> to be suffering from disease A

R_2: **If**   the patient exhibits symptom X **and**
         the patient exhibits symptom Y **and**
         the patient exhibits symptom Z
     then   the patient is <u>likely</u> to be suffering from disease B

Since the first two clauses of the antecedent of rule R_2 represent the consequent of rule R_1, R_2 could be reexpressed as:

R_2^*: **If**   the patient suffers from A **and**
           the patient exhibits symptom Z
       then   the patient is <u>likely</u> to be suffering from disease B

Other rules which do not constitute reversions of deductive relationships but could also be included in the knowledge-base are:

R_3: **If**   the patient suffers from A **and**
         the patient <u>does not</u> exhibit symptom Z
     then   the patient <u>is not likely</u> to be suffering from disease B

R_4: **If**   the patient suffers from A **and**
         the patient <u>does not</u> suffer from B
     then   the patient <u>could</u> be suffering from disease C

R_5: **If**   the patient suffers from A **and**
         the patient <u>does not</u> suffer from B
     then   the patient <u>could</u> be suffering from disease D

R_6: **If**   the patient suffers from A **and**
         the patient <u>does not</u> suffer from C **and**
         the patient <u>does not</u> suffer from D
     then   the patient <u>could</u> be suffering from disease B

The scheme of rules, as used above, does not adequately capture the given conceptual knowledge structure. The taxonomy of diseases is not explicitly represented through the rules; it is implicitly represented by repeating the same conditions in the antecedents of rules (e.g. the condition "the patient suffers from A" is present in the antecedent of every rule) and by including clauses that **restrict** competing hypotheses (e.g. "B is present", "C is present", and "D is present", i.e. the specialisations of A, represent a set of mutually exclusive hypotheses).

1.3 INFERENCE

Here we report on the forms of inference employed in the reasoning components of the majority of the publicized expert systems. The actual mechanization for implementing these two forms of inference depends on the scheme/s used for representing the knowledge captured within the systems. These schemes we divide into two categories. One category is the rule-based schemes and the other is a catch-all category (associative nets and frames).

1.3.1 Three Stages of Inquiry

In this section we introduce Peirce's (1931) terminology and his notions of the changing form of inference process of inquiry. The inference forms are characterized as having certain require-ments as a precondition for the validity of the inferences.

(1) We observe some puzzling phenomena and by abduction arrive at a certain hypothesis H.

(2) We deduce experimental consequences of H; experimental consequences are propositions of the form "If a procedure of a certain kind is carried out, a result of a certain kind will be observed".

(3) We carry out experiments from $(E_1 \ldots E_n)$ (finite). There are two cases:

    (i)   Suppose we find that, say, E_3 is false. Then we infer that H as it stands is false, though we may be able to give a modified version H^* from which E_3 does not follow.

    (ii)   Suppose $(E_1 \ldots E_n)$ are all true. Then we conclude by **induction** that either H, or some modified version of H, is the true explanation of the phenomenon.

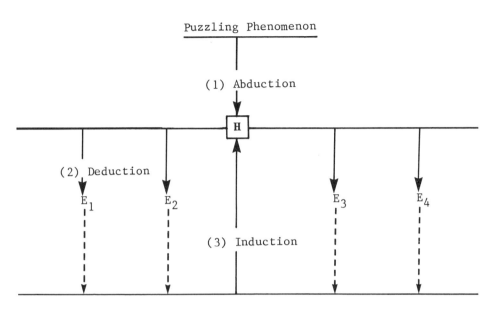

Requirements for abduction

(i)   The hypothesis must be such that some experimental consequences can be deduced from it ("pragmatic requirement").

(ii) The hypothesis must explain the puzzling phenomenon, hence it must be deducible from the hypothesis that such a phenomenon would occur.

(iii) A hypothesis which, if false, could be easily falsified is to be preferred.

(iv) An initially plausible hypothesis is to be preferred.

Requirements for deduction

(i) $(E_1 \ldots E_n)$ must follow by necessity from H.

Requirements for induction

(i)   "Fair sampling" requirements: these relate to the choice from all the possible experiments of those to be actually carried out.

(ii) "Predesignation": we must decide what hypothesis we are testing **before** making our observations.

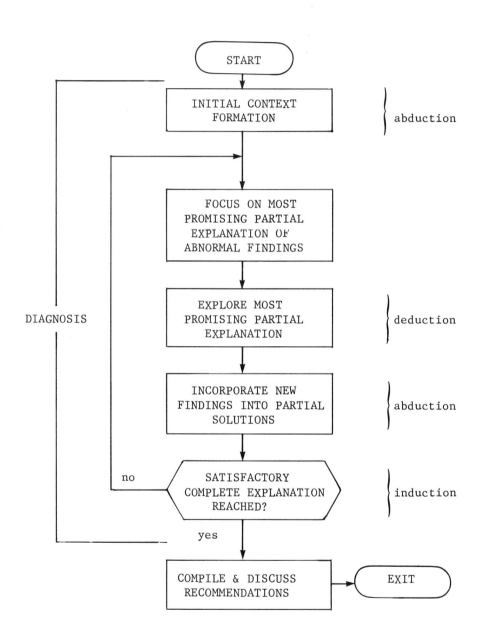

Fig. 1.5

1.3.2 Mechanization of Inference

Insight into the operations of expert systems may be gained by perceiving these operations as attempts to implement the inferences involved in the three stages of inquiry.

Rule-based schemes

In expert systems, rules chain together -- the consequent of one forms (part of) the antecedent of another -- to form inference networks. A rule-based system reasons by running along rule chains in a **forward** or a **backward** fashion, or a combination of the two. Forward-chaining (data or event-driven reasoning) is the mechanization of abductive inference and backward-chaining (goal or hypothesis-driven reasoning) is the mechanization of deductive inference.

Other schemes

Figure 1.5, shows, at an abstract level, the diagnostic cycle of reasoning that tends to be implemented when the knowledge representation scheme/s involved is not purely ruled-based. The various stages of the cycle are accompanied with an indication of the form of inference of which it is a mechanization.

1.4 PREVIEW

In this section we preview the systems described in this book. We indicate their function, their knowledge representation scheme/s, their notable features and how they exhibit the forms of inference in the three stages of inquiry.

MYCIN   **Function**: Diagnosis of, and recommendation of
               treatment for, antimicrobial infections. **Repre-**
sentation scheme: Rules. **Notable features**: The rule-based framework and its explanatory facilities. **Inference**: MYCIN diagnoses
patients by reasoning entirely in a backward, deductive, fashion.

PROSPECTOR   **Function**: Aids geologists in evaluating mineral
               sites for potential ore deposits. **Representation**
schemes: Rules and partitioned associative networks. **Notable**
features: The representation of rules in terms of partitioned
associative networks, as well as the extention of the strictly
rule-based framework to include taxonomies of concepts. **Infer-**
ence: PROSPECTOR diagnoses ore deposits by employing a mixture of

forward (abductive) and backward (deductive) reasoning.

PIP   **Function:** Simulates the behaviour of an expert
            nephrologist in taking the history of the present
illness of a patient with underlying renal disease. **Representation scheme:** Frames. **Notable features:** The formation of
contexts for problem solving via findings that act as abductive
triggers to hypotheses; the shifts of focus via links to
hypotheses with similar expectations (links to the differential
diagnosis). **Inference:** PIP abduces hypotheses via trigger links.
Also disorders complementary to hypothesised disorders are abduced
on the evidence of any of their associated typical expectations
(findings). Hypothesised disorders are explored by testing the
deductions (expected observations) inferrable on that hypothesis.

INTERNIST-I **Function:** Diagnosis of internal medicine.
             Representation scheme: Frame-like scheme. **Notable
features:** The formation of differential diagnoses; the synthesis
of differential diagnoses via the partitioning heuristic; the
information acquisition strategies. **Inference:** INTERNIST-I
abduces hypotheses using the differential diagnosis lists of the
"unexplained" manifestations. Competing hypotheses are investigated by testing the deductions drawable from them.

CADUCEUS   **Function:** Diagnosis of internal medicine.
             Representation scheme: Causal-taxonomical network.
Notable features: The restructuring of INTERNIST-I's factual
knowledge to permit two necessary, and synergistic, dimensions to
the diagnostic reasoning; the formation of the initial problem
contexts via special links (constrictors) from data to a point in
the diagnostic space. **Inference:** CADUCEUS abduces hypotheses from
constrictor observations and established pathological states. The
system differentiates competing hypotheses by testing the deductions drawable from them.

CASNET   **Function:** Long-term management of diseases whose
             mechanism is well known. **Representation scheme:**
Causal-associational network. **Notable features:** The division of
its factual knowledge into distinct planes of observations and
pathophysiological states; the separation between a belief measure
and a promise (of being correctly induced) measure for its state
hypotheses; the instantiation of disease pathways in the
pathophysiological plane. **Inference:** CASNET abduces state hypotheses from direct evidence and established causal, consequent
states; hypotheses are inductively supported by established

causal, antecedent states. Hypotheses are infirmed/ confirmed by testing the observations expected on them.

ABEL **Function:** Diagnosis of acid-base and electrolyte disorders. **Representation scheme:** Causal networks at different levels of abstraction. **Notable features:** The representation of a disease phenomenon at different levels of detail. The exploitation of the notion of causality in several ways: to organize the patient facts and disease hypotheses to deal with the effects of more than one disease present in a patient and to provide the basis of explanations. The capturing of the notions of adequacy and simplicity of a diagnostic possibility and hence providing the possibility of not requiring numeric belief measures as criteria for diagnostic reasoning. **Inference:** The initial hypotheses are abduced from electrolyte data using the acid-base nomograph. The construction of the diagnostic closures consisting of projecting backwards and forwards along the causal networks involves abducing hypotheses and deducing their expectations.

NEOMYCIN **Function:** To explicitly represent strategic knowledge and thus provide an efficient basis for teaching diagnostic reasoning and interpreting student behaviour. **Representation schemes:** Rules (object and meta) and frame-like structures. **Notable features:** The representation of its reasoning knowledge in an abstract fashion in terms of tasks and meta-rules; the generation of the strategy tree and the strategic explanations. **Inference:** NEOMYCIN abduces etiologies and important immediate state categories from the user observations. Hypotheses are tested through their expectations.

CRIB **Function:** Diagnosis of faults in computer hardware and software. **Representation scheme:** Frame-like scheme (implemented in a relational data base). **Notable features:** The use of predominantly heuristic rather than the standard algorithmic methods for automating the process of fault finding in machines; the information acquisition mechanism and primarily the system's learning mechanism. **Inference:** CRIB abduces faults/classes of faults from the observations conducted by the engineer and the symptoms assumed on the basis of previously hypothesised (more general) faults. The currently most localized hypothesised fault is tested by exploring the deductions drawable from the more specific faults subsumed under it. This cycle continues until a replacable/repairable unit can be induced to be faulty.

REFERENCES

Addis T.R. and **Johnson L.** (**1983**): "Knowledge for machines", in **The fifth generation computer project**, Pergamon Infotech.

Findler N.V. (ed.) (**1979**): **Associative networks**: representation and use of knowledge by computers, Academic Press, New York.

Hendrix G.G. (**1979**): "Encoding knowledge in partitioned networks", in Findler N.V. (ed.) **Associative networks**: representation and use of knowledge by computers, Academic Press, New York, pp. 51-92.

Johnson L. and **Hartley R.T.** (**1981**): "A short course in epistemology and knowledge engineering", **MCSG/TR 13**, Brunel University, Uxbridge UB8 3PH (UK). (Available from the authors on request.)

Johnson L. and **Keravnou E.T.** (**1983**): "The importance of the knowledge representation scheme in the performance of an expert consultant system", **MCSG/TR 29**, Brunel University, Uxbridge UB8 3PH, (UK). (Available from the authors on request.)

Minsky M. (**1975**): "A framework for representing knowledge", in Winston P.H. (ed.), **The psychology of computer vision**, New York, McGraw-Hill, pp. 211-277.

Nilsson N.J. (**1980**): **Principles of artificial intelligence**, Tioga Publishing Co., Palo Alto, California.

Peirce C.S. (**1931-58**): **Collected papers**, (eds.) Hartshorne C., Weiss P. and Burks A., 8 Vols. Cambridge, Mass.: Harvard University Press.
- "Three Types of reasoning", Vol. V pp. 94-111;
- "Kinds of Reasoning", Vol. I. pp. 28-31;
- "The Logic of Drawing History from Ancient Documents", Vol. VII, especially pp. 107-136;
- "Abduction, Deduction and Induction", Vol. VI. pp. 56-60;
- "Methods of Attaining Truth", pp. 391-432;
- "Deduction, Induction and Hypothesis", Vol. VI, pp. 372-386.

Chapter 2
MYCIN

Application area:	Medicine.
Principal researcher:	E. Shortliffe (H.P.P., Stanford University).
Function:	Diagnoses certain antimicrobial infections and recommends drug treatment.

MYCIN (Shortliffe **et al**, 1973; Shortliffe, 1976 and Davies **et al**, 1977) constitutes a novel attempt to incorporate the production rule methodology (Davis and King, 1977) from the general field of problem solving into the specific field of diagnostic problem solving. The name 'MYCIN' is taken from the common suffix shared by several of the drugs used in treatments e.g. clindamycin, erythromycin, kanamycin.

OVERVIEW

The knowledge-base is the program's store of task specific knowledge which deals with infectious disease diagnosis and therapy selection.

Knowledge-base

The knowledge is represented by domain-specific rules. Rules like the one below encode judgemental knowledge and are typically asserted with only a certain degree of confidence.

IF   the stain of the organism is gram negative **and**
       the morphology of the organism is rod **and**
       (the aerobicity of the organism is aerobic **or**
        the aerobicity of the organism is unknown)
THEN there is **suggestive** evidence that the class of
       the organism is enterobacteriaceae

Rule chaining

Assume that at some point in a consultation the system is trying to determine the "loci of infection" in the patient. The system retrieves all rules which make a conclusion about "the locus of infection" and invokes each one, in turn, evaluating each condition in the antecedent to see if it is satisfied. When evaluating a condition of a rule all rules which make a conclusion about that condition are retrieved and these are invoked, in turn, examining each condition, and so on. In this way rules chain together:

Let the goal be "loci of infection?" and let rule R_1 be in the set of rules invoked:

Rule R_1

IF   1) the culture site known **and**
        2) the specimen was collected this week **and**
        3) the organism causes a therapeutically
           significant disease
THEN   a locus of infection is definitely at the
        site of the culture

When evaluating the third condition of R_1, the subgoal is "therapeutically significant disease?". Rule R_2, say, is in the set of rules invoked:

Rule R_2

IF   1) the culture site is not normally sterile **and**
        2) the culture was collected by a sterile method **and**
        3) there is sufficient numbers of the organism
THEN   it suggests strongly, that the organism causes a
        therapeutically significant disease

Rules R_1 and R_2 may be chained together because in evaluating the antecedent of rule R_1, we come to evaluate condition (3) which sets up the subgoal of determining if the organism causes a therapeutically significant disease and rule R_2 is in the set of rules which conclude about this.

Diagnostic strategy

MYCIN "reasons backwards" from conclusions of rules to their conditions. The overall diagnostic goal is to determine if there are organisms, or classes of organisms, that require therapy. The system starts, therefore, by considering rules that conclude the existence of significant organisms in the patient, but the antecedent portion of each of those rules in turn sets up new subgoals. For example, at some point the system is likely to attempt to determine the loci of infection in the patient, and so on. The inference network grows as a tree structure in an AND/OR fashion -- because rules may have OR antecedent conditions.

When the question of whether a condition is satisfied is best determined by seeking laboratory data a question may be asked of the physician. Usually questions attempt to elicit the whole range of possible values for the given clinical parameter, and not just the value that is required to evaluate the rule condition. The physician's response is incorporated into the patient specific data base (context tree) and is thus available for evaluating conditions of other rules. After seeking laboratory data the system continues by invoking the appropriate rules. As information on a patient is particular, the inference network will be specific to the case under consideration, i.e. some rules will fail to be applicable because of the particular responses to questions. These same rules might, in another case, be applicable.

The search through the tree is depth-first because each antecedent condition is thoroughly explored in turn.

Inexact reasoning

Consider the case where the antecedent conditions of a rule, although in themselves assessed as certain, are only suggestive of the consequent rather than confirm it absolutely. In this case a number, less that 1 but greater than 0, is associated with the rule. The number is supplied by the domain expert as his/her subjective judgement as to the degree to which the antecedent is suggestive of the consequent. This is the usual case. These apriori 'certainty factors', however, are affected by situations that arise dynamically in a consultation. For example, if an antecedent of a rule has, in a particular case, only suggestive support then the number associated with the rule is modified to account for this further uncertainty. Consider the following rules (the apriori certainty factors are shown in brackets):

Rule R_3

IF 1) the site of the culture is the throat **and**
       2) the identity of the organism is streptococcus
THEN there is suggestive evidence (0.8) that the subtype
       of the organism is not group-D

Rule R_4

IF   1) the stain of the organism is gram positive **and**
       2) the morphology of the organism is coccus   **and**
       3) the growth conformation of the organism is chains
THEN   there is suggestive evidence (0.7) that the identity
       of the organism is streptococcus

Rule R_5

IF   brain-abscess is an infectious disease diagnosis
       for the patient
THEN   there is weakly suggestive evidence (0.2) that the
       identity of the organism is streptococcus

These rules chain together -- both R_4 and R_5 can provide
evidence ("the identity of the organism is streptococcus") which
contributes to the assessment of rule R_3. A simple function to
propagate the uncertainties of R_4 and R_5 into R_3 would be multiplication: if R_4 was confirmed the uncertainty associated with it
would be propagated into R_3 by the multiplication of 0.7 and 0.8
(=0.56); if R_5 was confirmed the uncertainty associated with it
would be propagated by the multiplication of 0.2 and 0.8 (=0.16).
In either case, as one would expect, the antecedent of R_3 is less
suggestive of its consequent than its apriori suggestiveness. In
other words mathematical functions propagate uncertainties over
inference and hence make the apriori certainty factors sensitive
to the context in which they are invoked. Because the reasoning
is inexact an exhaustive search strategy is used on the inference
network: should R_3 be confirmed on evidence provided by R_5, then
it would still seem wise to collect evidence in its favour because
it might be confirmed more highly on, for example, R_4.

Explanatory facilities

The invocation of a rule is taken as the fundamental action
of the system and this, together with the AND/OR tree as a
framework, accounts for sufficient of the system's operation to
make a trace of such actions an acceptable explanation. That is,
explanation is conceived in terms of the traversal of the case
specific AND/OR tree.

Suppose that the system asks "Has the organism been observed
in significant numbers?" and the physician responds "WHY", (Why is
it important to determine whether the organism has been observed
in significant numbers?). The system answers by displaying the
rules R_2 and R_1, in this order, thus showing that the item of
information was needed to determine if one of the conditions of

the subgoal "determine the loci of infection in the patient" is
satisfied. WHY is, therefore, answered by ascending the goal
tree.

The question "HOW" is interpreted as "How did you reach your
conclusion?", i.e. "How did you decide that the site of the
culture is a locus of infection in the patient?". The answer is
by showing the chaining. HOW is, therefore, answered by
descending the goal tree.

Therapy selection

If MYCIN determines that the patient has significant
organisms it goes on to determine the best therapy regime for that
patient. This objective is encoded in a goal rule "If there are
organisms that require therapy then determine the best therapy
regime". The specific therapy selection rules relate drugs to
organisms and patient specific data (context tree). The rules for
relating drugs to organisms are based upon statistical data as to
the drug's effectiveness in combating the organism.

2.1 STATICS

Rule-base

The domain is encoded in terms of about 450 production rules.
The rule format is given informally as: The antecedents (premises)
of rules are conjunctions of conditions; a condition is a
disjunction of simple assertions whose format is either
⟨predicate⟩ ⟨context⟩ ⟨parameter⟩ ⟨value⟩ or ⟨predicate⟩ ⟨context⟩
⟨parameter⟩. For example, the simple assertions "the identity of
organism-1 is streptococcus" and "the identity of organism-1 is
known" are, respectively, expressed as SAME(ORGANISM-1, IDENT,
STREPTOCOCCUS) and KNOWN(ORGANISM-1, IDENT). The consequents of
rules are simple assertions (for those rules encoding knowledge
manipulated by the 'diagnostic' process) or suggestions (for those
rules encoding knowledge manipulated by the therapy selection
process).

The rules used in the diagnostic process chain together and
implicitly define an AND/OR (inference) tree of assertions.
Informally the root assertion states that "there are organisms
that require therapy". This assertion, in fact, forms the
antecedent of the goal rule which connects the two components of
the system, namely diagnosis and therapy recommendations. The
rule, in full, states "If there are organisms that require therapy
then determine the best therapy regime".

There are meta-rules which function to control the invocation order of a specified subset of object-rules. (References to "rules" in subsequent sections imply object-rules).

Contexts and clinical parameters

MYCIN refers to the entities of importance in its knowledge domain as contexts. These are : persons, current-cultures, current-organisms, current-drugs, operations, operative-drugs, possible-therapies, prior-cultures, prior-drugs, prior-organisms. Rules are categorised in accordance with the context type for which they are most appropriately invoked. Each rule belongs to one and only one category.

Each context is associated with a number of clinical parameters (attributes), which fully describe that context. Clinical parameters are of three types:

Yes-no parameters - parameters that are either true or false;

Single-valued parameters - in general, they have a large number of possible values that are mutually exclusive, e.g the identity of an organism;

Multi-valued parameters - in general, they have a large number of possible values that need not be mutually exclusive, e.g the patient's drug allergies.

There are about eighty clinical parameters and each one is categorised in accordance with the context/s to which it applies. Some clinical parameters are primitive, i.e. values of them can be directly obtained from the physician, e.g the **age** of the person, whilst others are secondary and values of them depend on the values of primitive and/or secondary parameters, e.g the **identity** of an organism depends on the **stain, morphology** and **growth conformation** of the organism. Information relevant to each clinical parameter is kept in the knowledge-base as tables of properties of clinical parameters. Table 2.1 summarises the main contents of these structures.

Both hypotheses and pieces of evidence are statements asserting values of clinical parameters for the various contexts. Assertions that form antecedents of rules only -- assertions that can not been inferred from other assertions -- constitute the direct pieces of evidence. These are data directly asked of the user and represent the leaf nodes of the goal AND/OR tree. Assertions that figure in both consequents and antecedents of rules represent hypotheses at some point in the consultation, and 'circumstantial' evidence at another point. Such assertions would represent the intermediate nodes of the goal AND/OR tree. Finally assertions that form consequents of rules only, constitute ultimate hypotheses. Because the inference network is in fact a tree structure there would be just one such assertion.

Table 2.1: Properties of clinical parameters.

Property	Explanation
EXPECT:	Type of, and possible values for, the parameter.
LABDATA:	Indication as to whether the parameter is primitive or not.
LOOKAHEAD:	List of rules referencing the parameter in their antecedents.
UPDATED-BY:	List of rules inferring values for the parameter.
CONDITION:	Condition that prevents an unintelligent request for the value of the parameter (e.g. "Don't ask for an organism's sub-type if its genus is not known by the user").

2.2 DYNAMICS

The dynamics of the system have been abstracted in terms of the following rule, known as the **goal rule:**

RULE 092

IF   1) there is an organism which requires therapy and
       2) consideration has been given to the possible existence of additional organisms requiring therapy, even though they have not actually been recovered from any current cultures

THEN   Do the following:

1) compile the list of possible therapies which
   based upon sensitivity data, may be effective
   against the organisms requiring treatment, and

2) determine the best therapy recommendations from
   the compiled list

OTHERWISE:

indicate that the patient does not require therapy

The diagnostic strategy is a simple backward (goal-driven)
reasoning process; the topmost assertion that needs to be inferred
is the premise of the goal rule, "there are organisms that require
treatment". A **'context tree'** is dynamically constructed for the
patient. The context tree is a patient specific data base which
is the basis of the therapy selection process. Figure 2.1 gives
an example of a context tree; each node (context) in the tree may
be fully characterized by eliciting/inferring the value/s of every
clinical parameter defined for the given context. The therapy
selection process, abstracted in terms of the action part of the
goal rule, is invoked only if the diagnostic process has inferred
the presence of significant organisms, in the patient, that
require treatment.

2.2.1 Outlining the Overall Diagnostic Strategy
(Inference Engine)

In order to establish the top level assertion, the system
looks for rules whose conclusion is "there is an organism that
requires therapy". There are two such rules, rule 090 and rule
149. Suppose that rule 090 is evaluated first. This rule
informally states that "If the organism is significant then it
requires therapy". A subgoal would, therefore, be to determine
whether there is a significant organism. Hence, the system next
looks for rules whose conclusion is "there is a significant
organism", for example rule 038 and rule 042. Thus an AND/OR tree
is gradually built in a depth first fashion, by progressively
chaining backwards from each antecedent condition (figure 2.2
depicts the first few levels of the MYCIN AND/OR goal tree). When
the assertion to be established next constitutes a primitive piece
of data (labdata), then the relevant question is put forward to
the user, before attempting to infer it.

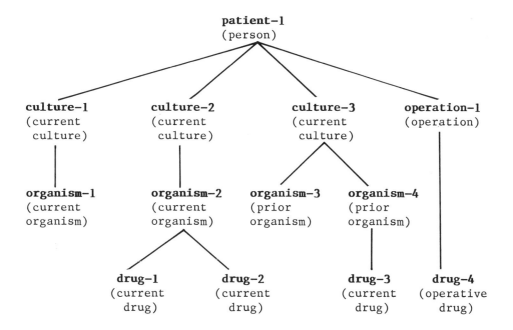

Fig. 2.1: Patient context tree.

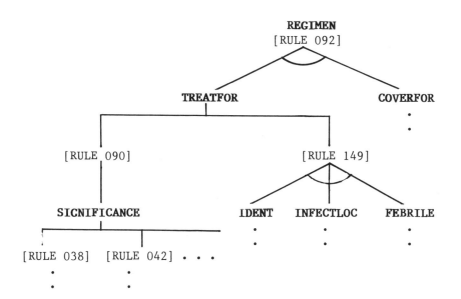

Fig. 2.2: MYCIN goal tree (adapted from Shortliffe, 1976).

The confidence in an assertion is updated as evidence is accumulated. When the confidence in the assertion is one of certainty the search stops. Usually evidence is only suggestive and hence the search is exhaustive. Only in the case where the overall confidence in some assertion is at least equal to a specified threshold value, does the assertion constitute evidence towards those assertions to which it is linked.

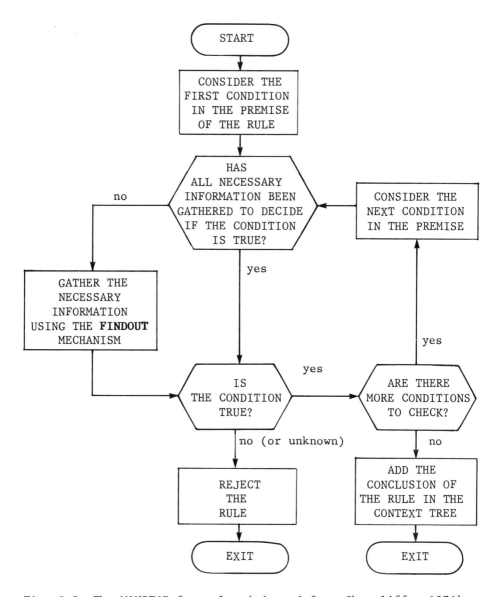

Fig. 2.3: The MONITOR for rules (adapted from Shortliffe, 1976).

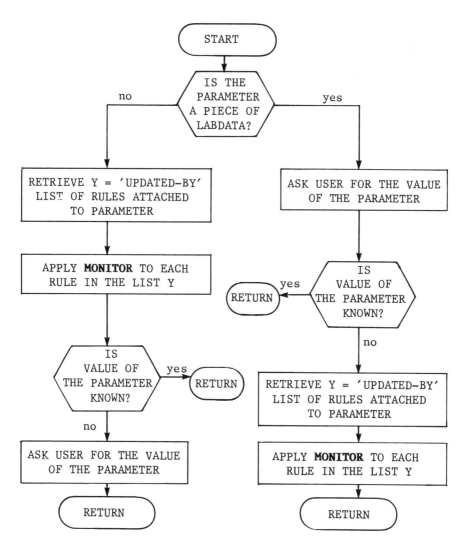

Fig. 2.4: The FINDOUT procedure (adapted from Shortliffe, 1976).

The diagnostic process is driven by two interrelated procedures, a MONITOR that analyses rules and a FINDOUT procedure which searches for data needed by the MONITOR. The flowcharts for MONITOR and FINDOUT are given in figures 2.3 and 2.4 respectively.

2.2.2 Assigning Uncertainties

MYCIN's model of inexact reasoning, outlined below, is

extensively discussed in Shortliffe and Buchanan (1975). It deals with two types of uncertainties:

      i)   uncertainties associated with rules
     ii) uncertainties associated with evidence

Uncertainties of rules

The uncertainty associated with a rule is given as an apriori certainty factor:

$$\underset{\text{(evidence)}}{\textcircled{E}} \xrightarrow{\;CF[H,E]\;} \underset{\text{(hypothesis)}}{\textcircled{H}} \quad ; \quad -1 <= CF[H,E] <= 1$$

$CF[H,E] > 0$ encodes the judgement that the hypothesis H is supported, by the definite occurrence of E, to that degree. $CF[H,E] < 0$ encodes the judgement that the hypothesis H is counter-indicated, by the definite occurrence of E, to degree $|CF[H,E]|$. $CF[H,E] = 0$ would mean that the evidence E is independent of the hypothesis (the knowledge-base contains no such rules).

The certainty factors do not obey the law of conditional probabilities that states $P(H/E) + P(\tilde{\ }H/E) = 1$. In fact,

$$CF[H,E] + CF[\tilde{\ }H,E] = 0, \quad \text{and hence,}$$
$$CF[H,E] = -CF[\tilde{\ }H,E] \qquad (1)$$

For example, when $CF[H,E] = 0.9$ (there is a certainty of 0.9 that **H is true** when E is true) we also have $CF[\tilde{\ }H,E] = -0.9$ (there is a certainty of 0.9 that **not H is false** when E is true). Thus, evidence can not both support a hypothesis and its negation.

Uncertainties of evidence

The rule antecedent conditions $C_1, \ldots, C_n$ have, in context, certainty factors $CF[C_i]$, and these determine the certainty factor $CF[E]$ of the rule antecedent. E's uncertainty is determined by taking the minimum uncertainty value over the C_i's. More generally, logical combinations of conditions can be taken; when AND the minimum CF value is taken as the CF value of the conjunct; when OR the maximum CF value is taken as the CF value of the disjunct. This policy is in accordance to the cannons of fuzzy set theory (Zadeh, 1965).

Referring to figure 2.5; let us say that the C_i's are

labdata, and the $CF[C_i]$'s are the confidence that the parameter
has a particular value. For example, the values of the stain,
morphology and growth pattern parameters of an organism, in
context, are expressed with uncertainty, and the minimum $CF[C_i]$
gives the $CF[E]$ in that context.

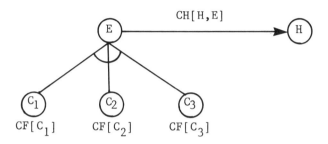

Fig. 2.5: Uncertainties of evidence.

Propagating uncertainties

The apriori certainty $CF[H,E]$ of a rule is modified by the
uncertainties in the context in which it is invoked:

$$CF[H,E'] = CF[E] * CF[H,E]$$

Combining evidence

The various propagated certainty factors of the rules
providing evidence E for a hypothesis H are combined to give the
overall certainty factor, $CF[H]$, of that hypothesis. If $CF[H] >= 0.2$ then the hypothesised assertion becomes evaluated in the
evidence of rules to which it is chained. $CF[H]$ is determined by
the following functions.

Each hypothesis has two measures associated with it, $MB[H]$
and $MD[H]$, both being assigned values in $[0,1]$. $MB[H]$ gives "the
measure of increased belief in the truthness of H" and $MD[H]$ gives
"the measure of increased disbelief in the truthness of H". The
confidence in the hypothesis is the algebraic sum of the current
evidence for and against it, i.e $CF[H] = MB[H] - MD[H]$. Thus, $-1 <= CF[H] <= 1$.

Initially $MB[H]$ and $MD[H]$ are set to zero. Each E_i such that
$CF[E_i] >= 0.2$ and $CF[H,E_i] > 0$ increments $MB[H]$, i.e. it provides
evidence in favour of the hypothesis. Each E_i such that $CF[E_i] >= 0.2$ and $CF[H,E_i] < 0$ increments $MD[H]$, i.e. it provides evidence
against the hypothesis.

For each E_i that provides evidence in favour of H we have:

$$MB[H,E_i] = CF[E_i] * CF[H,E_i]$$

and for each E_j that provides evidence against H we have:

$$MD[H,E_j] = CF[E_j] * |CF[H,E_j]|$$

By formula (1) above, when $MB[H,E_i] > 0$, $MD[H,E_i] = 0$, and when $MD[H,E_j] > 0$, $MB[H,E_j] = 0$. $MB[H,E_k] = MD[H,E_k] = 0$ when E_k is independent of H.

Two pieces of evidence, E_x and E_y, in favour of H are combined as follows:

$$MB[H,E_x\&E_y] = \begin{cases} 0 & \text{if } MD[H,E_x\&E_y] = 1 \\ MB[H,E_x] + MB[H,E_y] * (1-MB[H,E_x]) & \text{otherwise} \end{cases}$$

Similarly, two pieces of evidence, E_x and E_y, against H are combined as follows:

$$MD[H,E_x\&E_y] = \begin{cases} 0 & \text{if } MB[H,E_x\&E_y] = 1 \\ MD[H,E_x] + MD[H,E_y] * (1-MD[H,E_x]) & \text{otherwise} \end{cases}$$

Finally the total measures of belief and disbelief combine to give a certainty factor for H, CF[H], in [-1,1]:

$$CF[H] = MB[H] - MD[H]$$

Several hypotheses regarding a single-valued or multi-valued parameter may be simultaneously entertained by the system. In the case of a single-valued parameter the certainty factors of the competing hypotheses must not exceed 1. As soon as a hypothesis regarding a value for a single-valued parameter is proved to be true, all competing hypotheses are effectively disproved. This is not so for multi-valued parameters. In the case of yes-no parameters there is no need to consider "yes" and "no" as competing hypotheses since "yes" is "not no" (from equation (1) given earlier on, $CF["yes",E] = - CF["no",E]$).

Example 2.1

Suppose that MYCIN required the computation of the certainty of "the identity of organism-1 is streptococcus". The following rule is relevant:

IF   1) the stain of the organism is gram positive, and
         2) the morphology of the organism is coccus, and
         3) the growth conformation of the organism is chains
THEN there is suggestive evidence (0.7) that the identity
        of the organism is streptococcus

Let us say the context is as follows:

Organism-1:
          stain = grampos (1.0);
    morphology = coccus (0.6), rod (0.4);
growth pattern = chains (0.6), pairs (0.3), clumps (-0.8);
      •   •   •
      •   •   •

Let us say that the AND/OR tree that depicts the relevant portion of the inference network is given in figure 2.6. The shaded region represents the rule above, and the CF values of $C_{1.1}, \ldots, C_{1.3}$ are taken from the context.

$CF[H] = ?$ (H = "organism-1 is streptococcus")

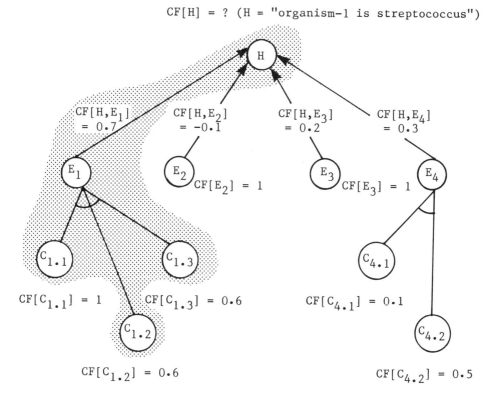

Fig. 2.6: AND/OR tree for hypothesis H.

We compute CF[H] as follows:

$$CF[E_1] = \min(1.0, 0.6, 0.6) = 0.6; \quad CF[H,E_1] = 0.7.$$

$$\therefore MB[H,E_1{}'] = 0.42.$$

$$CF[E_2] = 1; \quad CF[H,E_2] = -0.1.$$

$$\therefore MD[H,E_2] = 0.1.$$

$$CF[E_3] = 1; \quad CF[H,E_3] = 0.2.$$

$$\therefore MB[H,E_3] = 0.2.$$

$$\therefore MB[H,E_1{}'\&E_3] = 0.536.$$

$$CF[E_4] = \min(0.1, 0.5) = 0.1.$$

Since $CF[E_4] < 0.2$ the rule is not applied.

$$\therefore CF[H] = 0.536 - 0.1 = 0.436.$$

2.2.3 Focusing and Information Acquisition Strategies

MYCIN's focusing mechanism is superficial compared with the focusing mechanisms of other systems. Hypotheses are pursued in a predefined order yielding a dialogue directed solely by the system. The user is not allowed to redirect the chain of reasoning nor is the process at any point data-driven.

The main focusing strategy is the use of meta-rules (Davis and Buchanan, 1977 and Davies, 1980) for controlling the invocation of object-rules. Meta-rules are associated with clinical parameters; they are strategy rules which suggest the best approach to establishing the value/s of the associated parameter. For example, a meta-rule for the identity parameter of organisms states:

IF   1) the site of the culture is one of the non sterilesites **and**
        2) there are rules which mention in their premise a previous
           organism which may be the same as the current organism
THEN   it is definite (1.0) that each of them is not going to be
           useful

Thus meta-rules function to partition the relevant set of invoked rules (that provide evidence for the currently pursued assertion) in accordance to their degree of usefulness relative to the objective of establishing the assertion. Out of the most useful set of rules for a given hypothesised assertion, preference

is given to the rule that concludes the given assertion with the
highest certainty factor. Thus, rules for some hypothesised
assertion are not invoked in a predetermined order but their order
of invocation is dynamically determined during a consultation.

MYCIN attempts to display a more focused, methodical approach
to its diagnostic task by collecting all evidence about a clinical
parameter as it is encountered. For example, if the assertion
required to be established next by the system is that "the
identity of organism-1 is E.coli", the system not only attempts to
collect all evidence regarding the above assertion but also all
the evidence regarding the remaining statements asserting a
competing value for the identity of organism-1. This means that
all the rules that determine the value of the identity of an
organism, and not just the rules that determine whether the value
of the identity of an organism is E.Coli, are invoked at this
stage of the consultation. If the parameter in question is a
piece of labdata then the physician is asked about it before an
attempt to deduce its value/s is made. The physician, however, is
not asked the specific question "Is the value of the parameter so
and so?" but the general question "What is the value of the
parameter?". The response to the first question is a degree of
confidence whilst the answer to the second is a list of values
with associated degrees of confidence. For example, suppose that
the system wants to determine whether "the growth conformation of
the organism is chains". The question put forward to the
physician is the general question "Did the organism grow in
clumps, chains or pairs?". Thus, the system attempts to
elicit/deduce more information than is currently needed.

The focusing and information acquisition strategies of MYCIN
make its diagnostic strategy digress from a standard depth-first
search.

2.2.4 Therapy Selection Mechanism

The therapy selection mechanism deals with the action part of
the goal rule. Thus it consists of the following two-stage
process:

1)   Create a potential therapy list;

2)   Select the best drug or drugs from the list of
     possibilities.

Creation of potential therapy list

Drugs are selected on the basis of the identity of the

offending organisms. The likely identities for each significant organism are decided from the certainty factors of the relevant assertions held in the patient context tree. Each identity is given an item number.

The knowledge-base contains rules that provide sensitivity information for the various organisms known to the system. An example of such a rule is given below:

> **IF**   the identity of the organism is pseudomonas
>
> **THEN**   I recommend therapy chosen from among the following drugs:
>
> 1. colistin   (0.98)
> 2. polymyxin   (0.96)
> 3. gentamicin   (0.96)
> 4. carbenicillin   (0.96)
> 5. sulfisoxazole   (0.64)

The numbers associated with each drug are the probabilities that a pseudomonas isolated at Stanford Hospital will be sensitive (in vitro) to the indicated drug.

The potential therapy list for each item is given by the associated therapy rule.

Selecting preferred drugs

Firstly the best drug from the therapy list associated with each item is selected. The criteria used for the selection are:

1) probability that the organism is sensitive to the drug;
2) whether the drug has already been administered;
3) the relative efficacy of drugs that are otherwise equally supported by the criteria in (1) and (2) (the efficacy of a drug is based on whether the drug is bacteriostatic or bacteriocidal or its tendency to cause allergic sensitisation).

The next step is to attempt to minimize the set of best drugs by using heuristics like: "If drug-1 is the best choice for item-1 and the second best choice for item-2 then make drug-1 the best choice for item-2 as well". Facts like, "drug so and so can not be administered alone" or "not more than one drug from the same class should be administered" must also be taken into consideration. The manipulations of the drug lists outlined above are coded in LISP functions and not encoded as rules.

Finally the set of first-choice drugs is screened for contraindications against the patient specific information (e.g his/her particular drug sensitivities, age etc). The criteria for ruling out drugs are held in rules. If any first-choice drugs are contraindicated the therapy selection process is repeated without considering these. Once the best therapy list is recommended the physician may reject some of the recommended drugs. If this happens the process is again repeated using the remaining drugs.

2.3 FACILITIES

2.3.1 Explanatory Facilities

In the overview we characterized the operation of WHY and HOW questions as being ascent and descent of the inference network (see Shortliffe **et al,** 1975 and Davies **et al,** 1977). With the WHY? questions the user may specify the maximum number of tree levels which should be displayed. There is also the facility to ask "WHY didn't you" questions and the system gives rules that could have been used but failed. The user can specify a rule in this set and the system will respond with why the rule was not satisfied.

In addition to WHY, HOW and WHY didn't questions, the user can ask about the information that MYCIN has so far acquired (i.e. the system's belief that a given parameter has a given value), query the contents of production rules or request specific rules to be printed.

User tailored explanations

Wallis and Shortliffe (1982) introduce two measures, **complexity** and **importance,** which facilitate the generation of user tailored explanations. The currently reported outcome of this research is outlined below.

The two measures are assigned both to rules and 'concepts' (parameter values). Complexity of concepts denotes the level of knowledge required of the user in order to be in a position to grasp the concept and importance denotes how important the concept is for understanding the particular situation context in which the concept has been used. Complexity and importance of rules characterize, respectively, the level of knowledge required in order to see the inherent justification (logical basis) of the rule and the importance of the rule in the context (reasoning chain) in which it has been applied. In addition, for explanation

purposes, there are brief canned text-justifications of rules
which give the causal chains that explain the associations between
the antecedents and consequents of them.

The generation of explanations is to be guided by two user-
specific variables, **expertise** denoting the user's current level of
knowledge and **detail** denoting the level of detail desired by the
user when receiving explanations. A default value for the detail
measure can be computed from the specified expertise measure. The
chains of reasoning undertaken by the system are explained as
follows:

First, the intermediate concepts in a chain whose complexity
lies outside the expertise-detail range of the given user are left
out from the explanation of the reasoning chain -- unless the
concept's importance level exceeds a dynamic threshold, in which
case the importance measure overrides the complexity measure.

Second, if the complexity level of a rule is above the user
level of detail, then when the rule is selected on the basis of
its antecedent and consequent concepts as explained above, it must
be accompanied by its underlying canned text-justification. This
also applies for a rule, whose importance measure forced its
inclusion in an explanation of a reasoning chain containing it,
despite its complexity measure.

Example 2.2

Consider the following reasoning chain:

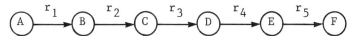

Figure 2.7 gives the complexities of concepts (A,...,F) and
rules (r_1,...,r_5) in relation to the user's expertise-detail
range. Supposing that the importance of concept C overrides its
complexity then for the given user the above reasoning chain would
be collapsed into A→C→D→F. Since rule r_3 (C→D) is too
complex for the user, the additional explanatory text associated
with the rule must also be provided.

2.3.2 Knowledge Acquisition Facilities

The explanatory facilities offer a means of knowledge
debugging. The expert has the option of stopping the program
after it has reached a conclusion and hence a chance to evaluate
the relevant parameters. Since facilities are provided for

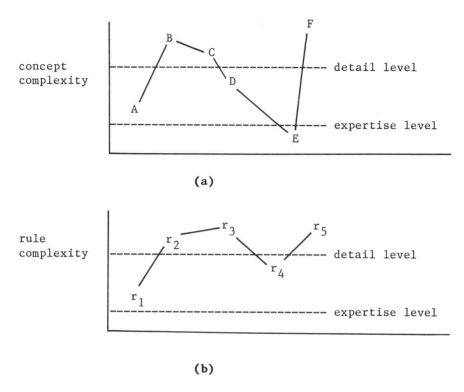

(a)

(b)

Fig. 2.7: Complexities of concepts and rules.

altering, adding or deleting rules, the explanatory facilities are also an appropriate environment for knowledge encoding ("acquisition in context").

Adding new rules to the collection or modifying existing rules could give rise to the following problems (see Shortliffe, 1976, pp. 213–215):

(i)   The addition of rule [a] A and B and C ⟶ D to a knowledge-base that already contains the rule [b] A and B ⟶ D would give rise to an instance of the problem of **subsumption**. Since rule [a] subsumes rule [b], in every situation where [a] can provide evidence for D, any evidence for D coming from [b] is redundant. To ensure that in no situation [b] counts redundantly towards D rule [b] must be converted into "A and B and ~ C ⟶ D".

(ii) The addition of rule A ⟶ B to a knowledge-base that already contains A ⟶ ~B would give rise to an instance of the problem of **single-rule inconsistencies.** The detection and resolution of such inconsistencies can be easily automated.

(iii) The addition of rule A →~D to a knowledge-base that already contains the chain A →B; B →C; C → D would give rise to an instance of the problem of **multiple-rule inconsistencies** (the new rule is inconsistent with a reasoning chain of old rules). The detection of such inconsistencies is a very difficult and time-consuming process.

The TEIRESIAS program (Davis, 1977, 1978, 1982 and 1983; Davis and Buchanan, 1977 and Davis and Lenat, 1982) is a sophisticated editor that aids a group of specified experts in updating MYCIN's knowledge-base. Modifications to the knowledge-base consist of adding new rules, modifying old rules or deleting old rules. TEIRESIAS takes care of the problems of subsumption and single-rule inconsistencies.

2.3.3 Teaching Facilities

GUIDON (Clancey, 1979a & 1979b and Davis, 1982) is a program that uses MYCIN to teach students. GUIDON's basic approach is to present the student with a case, supply additional information about the case when requested, and, by keeping track of the information requested, infer the problem solving approach being used by the student. If this approach appears inappropriate, GUIDON may interrupt the consultation to redirect the student's attention or to teach an appropriate rule from the knowledge-base.

2.3.4 System-building Facilities

EMYCIN (van Melle, 1979 and 1980), standing for "Essential **MYCIN**" is a system for building expert systems conforming to the essential MYCIN framework. The following systems have been built through EMYCIN:

PUFF (Aikins **et al**, 1982 and Kunz **et al**, 1978):

Domain of pulmonary disorders.
Performs interpretation of measurements from the pulmonary function laboratory.
Has 60 production rules.

HEADMED (Heiser et al, 1978):

Domain of clinical psychopharmacology.
Diagnoses a range of psychiatric disorders and can recommend drug
treatment if indicated.
Has 275 production rules.

SACON (Bennett and Engelmore, 1979):

Domain of structural analysis (engineering).
Provides advice to a structural engineer regarding the use of a
large structural analysis program called MARC that uses finite-
element analysis techniques to simulate the mechanical behaviour
of objects. The goal of the SACON program is to recommend an
analysis strategy to guide the MARC user in the choice of specific
input data, numerical methods and material properties.
Has 160 production rules.

ONCOCIN (Shortliffe et al, 1981):

Assists in the management of cancer patients on chemotherapy
protocols for forms of lymphoma.

CLOT (Bennett and Goldman, 1980):

Diagnoses disorders of the blood coagulation system.
Has 60 production rules.

DART (Bennett and Hollander, 1981):

Diagnoses software and hardware faults occurring within the tele-
processing (TP) subsystems for the IBM 370-class computers.
Has 190 production rules.

2.4 IMPLEMENTATION DETAILS

MYCIN is coded in INTERLISP. It is run under the TENEX
operating system that allocates 256 thousand virtual words of
memory (512 pages) to each user. For a MYCIN implementation,
approximately 320 pages are used for the INTERLISP system.
Approximately 100 pages (50K) are used for the compiled MYCIN
program. About 32 pages (16K) are required to hold the knowledge-
base. Approximately another 55 pages are required to contain the
clinical parameters, property tables and for working space. A
lengthy consultation requires approximately 20 minutes at a
computer terminal, including the time devoted to the optional use
of the explanation system.

REFERENCES

Aikins J.S., **Kunz J.C.**, **Shortliffe E.H.** and **Fallat R.J.** (1982):
"PUFF: an expert system for interpretation of pulmonary function
data", Stanford report **STAN-CS-82-931**.

Bennett J.S. and **Engelmore R.S.** (1979): "SACON: a knowledge-based
consultant for structural analysis", Proc. **IJCAI-79**, pp. 47-49.

Bennett J.S. and **Goldman D.** (1980): "CLOT: a knowledge-based
consultant for bleeding disorders", Technical Report, Computer
Science Department, Stanford University, Memo **HPP-80-7**.

Bennett J.S. and **Hollander C.R.** (1981): "DART: an expert system
for computer fault diagnosis", Proc. **IJCAI-81**, pp. 843-845.

Clancey W.J. (**1979a**): "Dialogue management for rule-based
tutorials", Proc. **IJCAI-79**, pp. 155-161.

Clancey W.J. (**1979b**): "Transfer of rule-based expertise through a
tutorial dialogue", Computer Science Doctoral Dissertation,
Stanford University, **STAN-CS-769**.

Davis R. (1977): "Interactive transfer of expertise: acquisition
of new inference rules", Proc. **IJCAI-77**, pp. 321-328.

Davis R. (1978): "Knowledge acquisition in rule-based systems.
Knowledge about representations as a basis for system construction
and maintainance", in Waterman D.A. and Hayes-Roth F. (eds.),
Pattern Directed Inference Systems, New York, Academic Press, pp.
99-134.

Davis R. (1980): "Meta-rules: reasoning about control", **Artificial
Intelligence, Vol. 15**, pp. 179-222.

Davis R. (1982): "Consultation, knowledge acquisition and
instruction: a case study", in Szolovits P. (ed.), **Artificial
Intelligence in Medicine**, AAAS Symp. Series, Boulder, Colo., West
View Press, pp. 57-77.

Davis R. (1983): "TEIRESIAS: experiments in communicating with a
knowledge-based system", in Sime M.E. and Coombs M.J. (eds.),
Designing for Human-Computer Communication, London: Academic
Press, pp. 87-137.

Davis R. and **Buchanan B.G.** (1977): "Meta-level knowledge: overview
and applications", Proc. **IJCAI-77**, pp. 920-927.

Davis R., Buchanan B.G. and Shortliffe E.H. (1977): "Production rules as a representation for a knowledge-based consultation program", **Artificial Intelligence, Vol. 8**, pp. 15-45.

Davis R. and King J. (1977): "An overview of production systems", **Machine Intelligence, Vol.8**, pp. 300-332.

Davis R. and Lenat D. (1982): **Knowledge-based systems in Artificial Intelligence**, New York: McGraw-Hill.

Heiser J.F., Brooks R.E. and Ballard J.P. (1978): "Progress report: a computerized psychopharmacology advisor", Proc. of the **11th Collegium Internationale Neuro-Psychopharmacologicum**, Vienna.

Kunz J., Fallat R., McClung D., Osborn J., Votteri B., Nii H., Aikins J., Fagan L. and Feigenbaum E. (1978): "A physiological rule-based system for interpreting pulmonary function test rules", Heuristic Programming Project Memo **HPP-78-19.**

Shortliffe E.H. (1976): **Computer-based medical consultations: MYCIN**, American Elsevier Publishing Co., Inc.

Shortliffe E.H., Axline S.G., Buchanan B.G., Merigan T.C. and Cohen S.N.(1973): "An artificial intelligence program to advise physicians regarding antimicrobial therapy", **Computers and Biomedical Research, Vol.6**, pp. 544-560.

Shortliffe E.H. and Buchanan B.G. (1975): "A model of inexact reasoning in medicine", **Mathematical Biosciences, Vol. 23**, pp. 351-379.

Shortliffe E.H., Davis R., Axline S.G., Buchanan B.G., Cordell Green C. and Cohen S.N. (1975): "Computer-based consultations in clinical therapeutics: explanation and rule acquisition capabilities of the MYCIN system", **Computers and Biomedical Research, Vol. 8**, pp. 303-320.

Shortliffe E.H., Scott A.C., Bischoff M.B., Campbell, vanMelle W. and Jacobs C.D. (1981): "ONCOCIN: an expert system for oncology protocol management", Proc. **IJCAI-81.**

van Melle W. (1979): "A domain-independent production-rule system for consultation programs", Proc. **IJCAI-79**, pp. 923-925.

van Melle W. (1980): "A domain-idependent production system for consultation programs", Computer Science Doctoral Dissertation, Stanford University.

Wallis J.W. and **Shortliffe E.H.** (1982): "Explanatory power of medical expert systems: studies in the representation of causal relationships for clinical consultations", **Meth. Inform. Med. Vol. 21**, pp. 127-136.

Zadeh L.A. (1965): "Fuzzy sets", **Information and Control, Vol. 8**, pp. 338-353.

Chapter 3
PROSPECTOR

Application area:	Geology.
Principal researchers:	P. Hart and R. Duda (SRI International).
Function:	Aids geologists in evaluating mineral sites for potential ore deposits.

OVERVIEW

PROSPECTOR (Gaschnig (1982), Duda **et al** (1979) and Reboh (1981)) is a geologists' aid to evaluating the favourability of an exploration site or region for ore deposits of particular types. The system accepts field observations from the user through an interactive dialogue and at the termination of the diagnostic process, produces a list of possible types of deposits present and their likelihood. If the likelihood, in the presence of ore deposits, is sufficiently high, then the system proceeds to determine the most favourable drilling sites.

The general knowledge associated with a particular class of deposit is coded in a set of production rules; uncertainties are associated with the rules. The 'inference network' resulting from the rule chainings, referred to as a **'model'**, is represented by the system as a partitioned associative network. Taxonomies of mineral types, rock types, physical forms and geological ages are also represented in the system's knowledge-base within the formalism of associative networks.

Case specific facts are instantiated assertions within the inference networks. These are gathered by asking the user questions. Initially a forward chaining is used and volunteered information is accepted from the user. After this initial stage the system switches to backward chaining and asks for information, but the user may still volunteer information and this may result in a new current goal hypothesis being chosen to investigate.

The hypothesis, asserting the presence of some deposit class, having the highest current probability is always selected as the goal. Every untried rule having the current hypothesis as a consequent is scored according to how heavily it may influence it (the current hypothesis) and the highest scoring rule is selected. In this way a chain is linked.

3.1 STATICS

PROSPECTOR diverts from being a strictly rule-based system, by allowing for taxonomies of concepts to be explicitly represented in its knowledge-base. (The performance of the system depends critically on this extention (see Reboh, 1981, ch. V)).

The conceptual components of PROSPECTOR's knowledge-base, namely the taxonomies of concepts and the inference networks (each network modelling some class of deposits) are represented within the scheme of partitioned associative networks (Hendrix, 1979).

3.1.1 Taxonomies

The knowledge-base contains taxonomies of rock types, minerals, physical forms and geological ages (figure 3.1 gives a portion of the materials taxonomy).

Taxonomies are represented as hierarchical tree structures where the nodes are simple concepts of the domain connected by arcs indicating the element (**e**) and subset (**s**) relationships between these concepts.

Each node X in the hierarchical structure is said to be a **restriction** of its parent nodes or of any node occurring on a chain of outgoing "e" and "s" arcs from X. Similarly, X is a **generalisation** of any descendant node occurring on any chain of incoming "e" and "s" arcs to it.

For a more precise encoding of taxonomies, the standard set theory notions of set membership and set inclusion ("e" and "s" arcs respectively), are supplemented by the more restrictive notions of disjoint subsets (**ds**) and distinct elements (**de**). A "ds" arc from a node X to a node Z indicates that X is a subset of Z and that X is disjoint from any other set Y with an outgoing "ds" arc to Z. Similarly, arcs labelled "de" indicate that each of two or more nodes denotes a different element of a set.

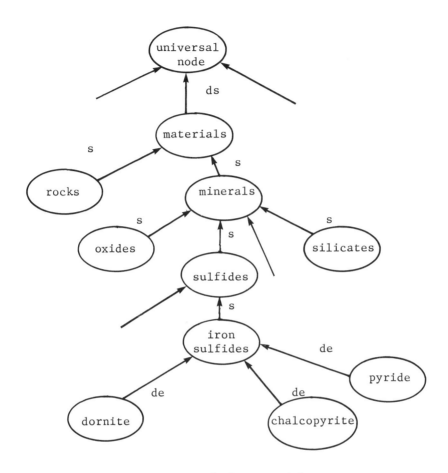

Fig. 3.1: Part of the materials taxonomy
(adapted from Reboh, 1981).

Table 2.1: Relations

binary	n-ary (n>2)
age-of	altered-to
comp-of (composition)	components-of
form-of	contained-in
grain-size-of	relative-age-of
loc-of (location)	
size-of	
texture-of	

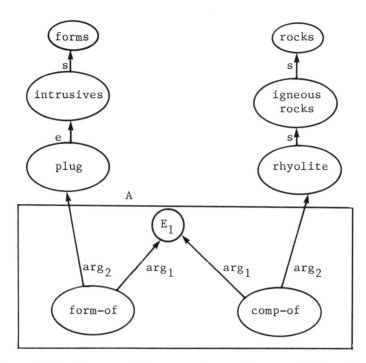

Nodes within the partition can be used as **variables** (e.g. E_1 above) that can be bound to constants outside the partition.

Fig. 3.2: Associative network representation of assertion "a rhyolite plug is present".

3.1.2 Inference Networks

The rules coding the general knowledge associated with some class of deposit chain to implicitly define an inference network of assertions, referred to as the **model** of the deposit class. (The knowledge-base currently contains fifteen such prospect-scale models; each model consists of more than 150 rules and over 200 assertions.) Rules are assigned numeric values (sufficiency, S, and necessity, N, factors) that specify the strength of the relevant inferences.

Each inference network indicates how the observations (field evidence), associated with the given deposit class, relate to intermediate geological hypotheses which eventually relate to the ultimate hypothesis. This hypothesis asserts the presence of the relevant deposit class (inference networks are therefore hierarchical in nature).

Before we give the partitioned associative network representation of an inference network (see Duda **et al**, 1978) we make a
small digression to discuss how assertions (forming the antecedents and consequents of rules) are represented within this
scheme.

Assertions

Assertions are represented in terms of partitions, enclosing
nodes representing physical entities, processes, places, and
relations and arcs representing arguments. Relations are represented as nodes rather than arcs to permit for n-ary relations
where n>2 as well -- Table 3.1 gives some binary and n-ary
relations (n>2) used in PROSPECTOR for expressing assertions.

Figure 3.2 gives the network representation of the assertion,
A, "a rhyolite plug is present ". This representation is
obtained by analysing A into the three primitive assertions:

(A1) there exists a physical entity E_1;
(A2) the composition of E_1 is "rhyolite";
(A3) the form of E_1 is "plug".

Different assertions can also be combined logically to form a
single, compound assertion. The simpler elements are combined by
means of the logical connectives AND, OR and NOT. The network
representation of the compound assertion (A), "A_1 AND A_2 AND A_3"
is given in figure 3.3.

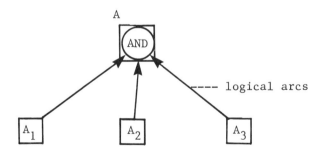

Fig 3.3: Associative network representation of the compound
assertion "A_1 AND A_2 AND A_3".

Inference rules

The associative network representation of a PROSPECTOR rule
is given in figure 3.4. A rule is represented in terms of a rule-

arc (**plausible link**) linking the partition enclosing the network representation of the rule antecedent (evidence statement) to the partition enclosing the network representation of the consequent of the rule (hypothesis statement).

Fig. 3.4: Associative network representation of a PROSPECTOR inference rule.

Rules can interconnect in various ways -- through "chains" where the consequent of one rule forms (part of) the antecedent of another, through sharing the same consequent or through sharing (part of) their antecedents. Rules can also be linked implicitly through the element-subset "chains" e.g. when the antecedent of one rule is a restriction of the antecedent of another rule having the same consequent.

A simple inference network is illustrated in figure 3.5. The inference network representing the model of some ore-deposit is a hierarchical structure of assertions, linked through rule and logical arcs, as depicted in the figure. The principal or top-level assertion in an inference network is the assertion that the available evidence matches that particular model. It is therefore the model's goal assertion. The terminal or "leaf" assertions correspond to field evidence asked of the user. Every assertion is assigned an initial (prior) probability.

Apart from the plausible and logical links discussed above, figure 3.5 depicts a third type of links, the **contextual links.** Contextual links are used to specify the order in which the relevant assertions are to be considered, when arbitrary ordering is meaningless (i.e. contexts are used to express conditions that must be established before an assertion can be used in the reasoning process), or in order to specify that one assertion is geologically significant only if another assertion has already been established. The relationships represented through the contextual links are not categorical in nature (see Reboh, 1981).

3.2 DYNAMICS

Assertions corresponding to terminal nodes of the inference

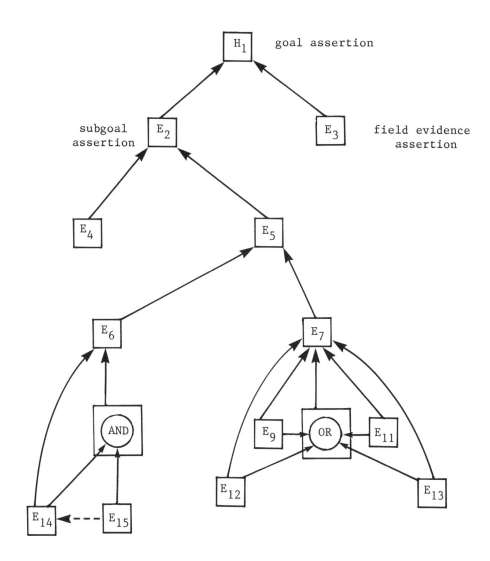

KEY:

⟶▶: plausible (rule) arc; ⟶▶: logical arc;
- - - -▶: contextual arc.

Fig. 3.5: A simple inference network.

networks constitute direct (field) evidence, assertions
corresponding to intermediate nodes constitute both intermediate
geological hypotheses and 'circumstantial' evidence for higher
level assertions and assertions corresponding to the topmost nodes
constitute the ultimate (goal) hypotheses. The objective of the
system is to determine the ore deposits likely to be present in
the given situation context and recommend drilling sites. The
first stage of a consultation (diagnosis of ore deposits) is an
on-line process engaging in a mixed-initiative dialogue with the
user. The second stage of a consultation (recommending drilling
sites) is a time consuming process that does not require the
participation of the user and it is thus executed in an off-line
fashion.

The basic dynamics of the system (diagnosis of ore deposits)
can be abstracted in terms of the assertion status transition
diagram of figure 3.6. Initially the validity of every assertion
in the given situation context is **undetermined.** The validity of
assertions corresponding to pieces of field evidence is determined
by the user whilst the validity of intermediate and goal
assertions is determined by the system (through propagating the
field evidence onto the inference networks). An assertion is
therefore **determined** when its degree of truthness (posterior
probability) in the given situation has been fully established.
Thus the determined status can be refined into a whole sequence of
status ranging from **confirmed** to **denied.** An undetermined
assertion is **abduced** on the basis of direct/circumstantial
evidence of any strength (forward chaining from direct evidence).
An assertion is **pursued** (i.e. the system attempts to establish its
validity) if it is **deduced** on the basis of the currently pursued
hypothesis (backward chaining) and it constitutes the most
'influential' potential piece of evidence for the currently
pursued hypothesis. At every diagnostic stage the first assertion
to be pursued is the currently most likely (abduced) goal
assertion. The pursuing of an assertion terminates when all the
relevant evidence has been explored; when this happens the truth
status of the assertion in the given context has been fully
determined.

3.2.1 Outlining the Overall Diagnostic Strategy (Inference Engine)

At the beginning of a consultation the user can volunteer any
observations (field evidence) he/she considers significant. Such
observations are used to update the prior probabilities of the
goal assertions by forward chaining from the terminal assertions
thus instantiated.

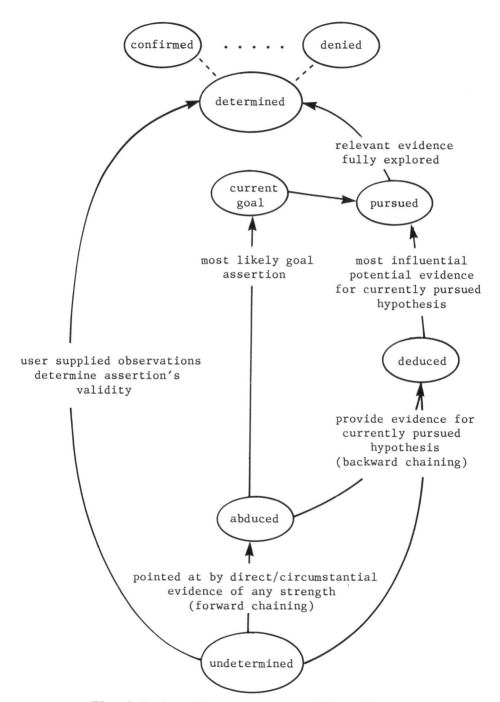

Fig. 3.6: Assertion status transition diagram.

At every subsequent stage of the diagnostic process the system attempts to establish the most likely goal assertion, i.e. to establish the presence of the relevant deposit class. (If no initial information has been volunteered by the user then the models are pursued in an order predetermined by the prior probabilities of their goal assertions.) A goal assertion is pursued by backward chaining: all the rules that conclude the assertion are determined and the antecedent of the rule judged as the one to influence the current probability of the pursued assertion to the greater extent becomes the new hypothesis to be pursued. Thus a rule chain, leading to a terminal assertion is gradually linked. When the hypothesis to be pursued next corresponds to a piece of field evidence then the user is queried about it. The effect of the user response is then propagated immediately along the inference chains until the probability of every goal assertion, reachable from the given terminal assertion, has been updated. If by the end of this propagation the current goal assertion is still the most likely one then backchaining resumes from where it had been interrupted. Otherwise back-chaining starts from the new most likely goal assertion. This is not possible in MYCIN where certainty factors (not probabilities) are propagated one level only.

At any point during the diagnostic process the user can interrupt the system to suggest an alternative hypothesis to the one the system is currently pursuing or to enter any additional observations considered relevant. The effect of any new observations is forward chained and this may result in a new current goal hypothesis being investigated.

Connections between the different models function to promote the investigation of those models related to already established models. The diagnostic process terminates when every sufficiently promising model has been fully investigated.

Handling of user observations (forward chaining)

User observations are entered as simple declarative sentences accompanied with "certainty factors" between −5 and 5, denoting the user's degree of belief or disbelief in the sentences' truthness. (Certainty factors are mapped into posterior proba-bilities by the relationship depicted in figure 3.7.). The system converts such sentences into their associative network represent-ations and matches them against the knowledge-base assertions. A user observation which does not match with any assertion is ignored. A user observation which yields matches with assertions initiates further processing.

The match between a sentence referring to a user observation and a knowledge-base assertion is classified as one of:

(i)   **identity,**

(ii) **restriction** (or **generalisation**) (the taxonomies of concepts
      play an important role in determining whether a sentence is
      a generalisation/ restriction of another sentence), or

(iii) **overlapping** (i.e. a match between some of the primitive
      component sentences of the two assertions occurs).

A rule whose antecedent is either identical to, or a
generalisation of, some user observation can be applied (i.e.
forward chaining takes place). However this is not so for a rule
whose antecedent is either a restriction of, or overlapps with,
some user observation. In such cases additional information needs
to be acquired from the user for the relevant rules to be rendered
applicable.

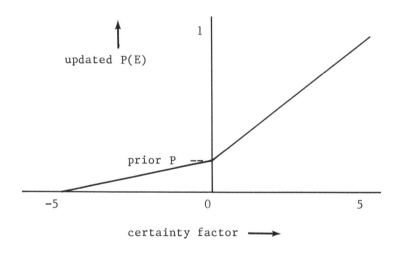

Fig 3.7: Relation between certainty and probability
(adapted from Reboh, 1981).

The designers of PROSPECTOR believe that by permitting the
explicit representation of conceptual structures like taxonomies
there is a basis for building mechanisms for detecting inconsistencies in the evidence provided by the user. For example, the
user's degree of confidence in the validity of some assertion
should not exceed his/her degree of confidence in its generali-
sations. This can be checked if there is an explicit
representation of the relationship between the assertions.

3.2.2 Assigning Posterior Probabilities to Assertions

PROSPECTOR's model of inexact reasoning is probabilistic. It dictates how to derive the degree of belief in a conclusion from the degrees of belief in the antecedents, on a base drawn from Bayes' theorem. A concise description of the model is given in Duda **et al** (1976). The basic concepts of the model are summarised below.

Types of uncertainties

The diagnostic process deals with two types of uncertainties:

   i) uncertainties associated with the rules, and
   ii) uncertainties associated with the evidence.

A PROSPECTOR production (inference) rule is of the format:

   If E then (to degree S, N) H

S (sufficiency factor) and N (necessity factor) are numeric values that specify the strength of association between E and H. They are calculated from the following formulae:

$$S = \frac{P(E/H)}{P(E/\sim H)}$$

S represents the degree of sufficiency of E in establishing H. If S approaches infinity then E is sufficient to establish H.

$$N = \frac{P(\sim E/H)}{P(\sim E/\sim H)}$$

N represents the degree of necessity of E in establishing H. If N approaches zero then E is necessary for H. The above conditional probabilities are not objective, i.e. they are not calculated from statistical samples, but they are referred to as **subjective probabilities** extracted from the experts.

The inference networks' assertions are preassigned subjective a priori probabilities by the experts. During a consultation these probabilities are updated using the user observations (such posterior probabilities are eventually converted into certainty factors using the relationship depicted in figure 3.7). In general the user cannot specify that a piece of evidence is definitely present or absent, but instead, he/she can denote a degree of confidence that the evidence sought is actually present.

Combining types of uncertainties

From Bayes' theorem,

$$P(H/E) = \frac{P(E/H) \; P(H)}{P(E)}$$

the following two equations can be deduced:

$$O(H/E) = \frac{P(H/E)}{P(\sim H/E)} = \frac{P(E/H)}{P(E/\sim H)} * \frac{P(H)}{P(\sim H)} = S * O(H) \quad (1)$$

$$O(H/\sim E) = \frac{P(H/\sim E)}{P(\sim H/\sim E)} = \frac{P(\sim E/H)}{P(\sim E/\sim H)} * \frac{P(H)}{P(\sim H)} = N * O(H) \quad (2)$$

where: O(H)   denotes the prior odds on H being true,
    O(H/E)   denotes the posterior odds on H being true given
             that E is observed to be present, and
    O(H/~E) denotes the posterior odds on H being true given
             that E is observed to be absent.

Probabilities can be recovered from their odds by the simple formula:

$$P = \frac{O}{O+1}$$

Equation (1) is used to update the probability on H when E is observed to be present. This updating is based on S, the sufficiency factor. Equation (2) is used to update the probability on H when E is observed to be absent. This updating takes into consideration N, the necessity factor. Below we explain how these updating formulae can be extended to accommodate uncertainty in the evidence.

Let E' denote the observations that cause the user to suspect the presence of the evidence E. Formally the posterior probability P(H/E') is given by:

$$P(H/E') = P(H,E/E') + P(H,\sim E/E')$$
$$= P(H/E,E') \; P(E/E') + P(H/\sim E,E') \; P(\sim E/E')$$

Duda **et al** (1976) says that by making the "reasonable" assumption that "if we know E to be present (or absent) then the observations E' relevant to E provide no further information about H " (i.e. by assuming that P(H/E,E') = P(H/E) and P(H/~E,E') = P(H/~E)) the above equation becomes

$$P(H/E') = P(H/E) \; P(E/E') + P(H/\sim E) \; (1-P(E/E'))$$

Thus under the above assumption, P(H/E′) is a linear function of P(E/E′), with P(H/E′) = P(H/~E) when P(E/E′) = 0 and P(H/E′) = P(H/E) when P(E/E′) = 1.

Because the probability values P(H/E), P(H/~E), P(H) (prior) and P(E) (prior) are all subjectively obtained, they may often be inconsistent with the theoretically expected values. In particular, it must be made certain that when nothing is known about E, i.e. when P(E/E′) = P(E), the interpolation function should leave H at its prior probability, yielding the theoretically expected value P(H/E′) = P(H).

P(H/E′) is therefore chosen to be a piecewise linear function of P(E/E′) so that the desired values for P(H/E′) are obtained at the three fixed points P(E/E′) = 0 , P(E), and 1. The resulting function is shown graphically in figure 3.8. The desired posterior odds O(H/E′) can thus be computed, and hence an estimate of the effective odds multiplier, λ, can be obtained as

$$\lambda = \frac{O(H/E')}{O(H)}$$

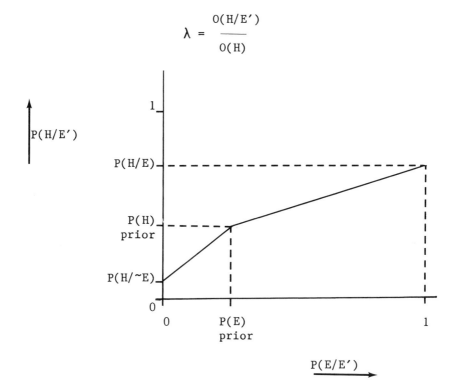

Fig. 3.8: Consequent probability as a function
of the antecedent probability
(adapted from Reboh, 1981).

Example 3.1

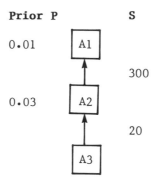

Fig. 3.9: Propagating uncertainties.

Referring to figure 3.9, suppose that A3 (field evidence) has been observed with certainty. This finding will be propagated onto the rule chain to update the probabilities on assertions A2 and A1.

Since A3 is definitely present equation (1) is used:

$$O(A2/A3) = 20 \text{ x } (0.03 / (1 - 0.03)) = 0.619.$$
$$\therefore P(A2/A3) = 0.619 / (1 + 0.619) = 0.39.$$

Since A2 is not established with certainty the odds multiplier corresponding to A2′ must be modified as:

$$\lambda_{A2'} = 300 \text{ x } ((0.382 - 0.03) / (1 - 0.03)) = 108.9.$$

and therefore,

$$O(A1/A2') = 108.9 \text{ x } (0.01 / (1 - 0.01)) = 1.1.$$
$$\therefore P(A1/A2') = 1.1 / (1 + 1.1) = 0.524.$$

Combining evidence towards the same hypothesis

It often happens that several pieces of evidence bear on the same hypothesis. In other words there are rules with the same consequent condition but different antecedent conditions. This is illustrated in figure 3.10.

By assuming that the E_i's are conditionally independent under both H and ~H the following formula can be used for updating the probability on H:

$$O(H/E_1, E_2, \ldots, E_n) = \left[\prod_{i=1}^{n} \lambda_i \right] O(H)$$

where: λ_i denotes the odds multiplier corresponding to E_i.

The assumption of independence mentioned above is challenged in Pednault **et al** (1981).

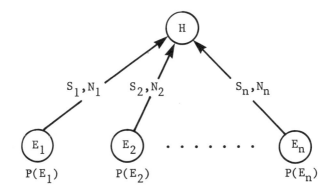

If some piece of evidence represents a logical combination of other pieces of evidence then its probability can be obtained from the probabilities of its components: AND performs a minimization operation, OR does a maximization and NOT a complement operation.

Fig. 3.9: Multiple pieces of evidence.

3.2.3 Focusing and Information Acquisition Strategies

PROSPECTOR's diagnostic process involves forward (data-driven) reasoning; the user is allowed to volunteer information whenever he/she wishes. Any information submitted at the beginning of a consultation is used to constrain the system's focus of attention to the most likely deposit classes (models). At every diagnostic stage the most likely model is pursued. Additional user information could result in shifting the system's focus of attention to a new model for investigation. The taxonomies of concepts play an important role in information acquisition, since they enable the system to see that although a user observation does not yield a complete match with a rule antecedent , still it is not a completely irrelevant observation (e.g. it is a generalisation of a rule antecedent) and, therefore, it should be further explored before it is disregarded.

When the diagnostic process is reasoning backwards from the model under investigation the following focusing heuristics are employed:

i)   If the currently pursued hypothesis is a disjunction/ conjunction of assertions then the component assertion with the largest/smallest current probability is selected as the next hypothesis to be pursued. The effectiveness of this focusing heuristic are inherent in the interpretations assigned to the AND and OR connectives; in each case the selected component assertion is the one most likely to influence the probability of the complex assertion.

ii) Every untried rule having the currently pursued hypothesis as a consequent is scored according to how heavily it may influence the hypothesis' likelihood. The highest scoring rule is selected. The heuristic scoring function employed (to dynamically order the applicable rules) is:

$$Sc(E_i) = |\log(S_i/\lambda_i)|P(E_i/E_i') + |\log(\lambda_i/N_i)|(1-P(E_i/E_i'))$$

where: E_i is the antecedent of the ith applicable rule,
S_i/N_i are the sufficiency/necessity factors assigned to the rule, and
λ_i denotes the odds multiplier corresponding to E_i.

When the antecedents are a priori unlikely, the heuristic favours the rules with small necessity factors, and as they become more likely it favours the rules with large sufficiency factors.

Finally the contextual links are a constituent of PROSPECTOR's focusing and information acquisition mechanism. Those contextual links that establish the ordering of pursuing hypotheses, ensure an intelligible sequence of information acquisition from the user, and those links that specify whether some hypothesis is geologically significant only if another hypothesis has already been established ensure a promising sequence in the pursuing of hypotheses (these latter type contextual links are analogous to MYCIN'S meta-rules).

3.2.4   Drilling Sites Selection Mechanism

The first stage of the consultation consists of determining whether there is sufficient evidence pointing to the presence of ore deposits. If this is so PROSPECTOR proceeds to the second stage of the consultation, i.e. the recommendation of drilling sites. During this stage the system derives its inputs from digitized maps of geological characteristics (these maps are obtained interactively from the user during the previous stage of the consultation and are saved on external data files), and produces as output a colour-coded graphical display of the favourability of each cell on a grid corresponding to the input map. Hence this stage of the consultation is more appropriately run in batch mode rather than interactive mode. Konolige (1979) explains how the drilling site selection process can be used in batch mode.

3.3 FACILITIES

3.3.1 Explanatory Facilities

PROSPECTOR provides a number of explanatory facilities, most important amongst these being the WHY (why a given question was asked) and HOW (how a given conclusion was reached) explanation mechanism drawn from MYCIN. Further the user can ask for the geological rationale for including a question in a model.

3.3.2 Knowledge Acquisition Facilities

The Knowledge Acquisition System (KAS) (Reboh, 1981) guides the experts during the several stages of elaboration, refinement and revision required to construct prospect-scale models conforming to the format described in section 3.1.2. The system is in a position to detect any inconsistencies in the knowledge submitted by the expert. A recent addition to the system, which is still under investigation, is the ability to merge in the same knowledge-base conflicting expertise extracted from different experts (Reboh, 1983).

REFERENCES

Duda R., Hart P. and **Nilsson N.** (1976): "Subjective Bayesian methods for rule-based inference systems", Proc. National Computer Conference (**AFIPS Conference Proceedings**), **Vol. 45**, pp. 1075-1082.

Duda R., Hart P., Nilsson N. and **Sutherland G.L.** (1978): "Semantic network representations in rule-based inference systems", in Waterman D.A. and Hayes-Roth F. (eds.) , **Pattern Directed Inference Systems,** New York, Academic Press, pp. 203-222.

Duda R., Gaschnig J. and Hart P. (1979): "Model design in the PROSPECTOR consultant system for mineral exploration", in Michie D. (ed.), **Expert Systems in the Microelectronic age**, Edinburgh University Press, pp. 153-167.

Gaschnig J. (1982): "PROSPECTOR: an expert system for mineral exploration", in Michie D. (ed.) , **Introductory readings in Expert Systems**, Gordon and Breach Science Publishers, Inc., pp. 47-64.

Hendrix G.G. (1979): "Encoding knowledge in partitioned networks", in Findler N.V. (ed.), **Associative Networks: representation and use of knowledge by computers**, Academic Press, New York, pp. 51-92.

Konolige K. (1979): "An inference net compiler for the PROSPECTOR rule-based consultation system", Proc. IJCAI-79, pp. 487-489.

Pednault E.P.D., Zucker S.W. and Muresan L.V. (1981): "On the independence assumption underlying subjective Bayesian updating", **Artificial Intelligence, Vol. 16**, pp. 213-222.

Reboh R. (1981): "Knowledge engineering techniques and tools for expert systems", Linkoping Studies in Science and Technology, Dissertations NO. 71, Software Systems Research Center, Linkoping University, S-581 83 Linkoping, Sweden.

Reboh R. (1983): "Extracting useful advice from conflicting expertise", Proc. **IJCAI-83**, pp. 145-150.

Chapter 4
PIP

Application area:	Medicine.
Principle researchers:	S.G. Pauker and P. Szolovits (Massachusetts Institute of Technology).
Function:	Simulates the behaviour of an expert nephrologist in taking the history of the present illness of a patient with underlying renal disease.

OVERVIEW

Perhaps the most appropriate way to give an overview of the Present Illness Program (PIP) (Pauker **et al,** 1976 and Szolovits and Pauker 1976, 1978) is to adumbrate Pauker and Szolovits' (1977) model of expert's clinical behaviour in their domain of expertise and then to adumbrate the form and operation of the computer model.

Clinical model

Presented with a complaint the physician's initial purpose is to develop reasonable (and robust) hypotheses from the patient's complaint and the data acquired during the interview. This process is the formation of the context in which the physician will formulate deeper questions and make decisions regarding diagnostic tests, etc.

An expert physician can make an initial guess as to the nature of the problem with very little information (see Elstein **et al,** 1978). When presented with certain key findings ("triggers") he/she rapidly establishes a context for further exploration. As additional information is gathered, it is evaluated with respect to the current working hypotheses, and these working hypotheses are revised. The expert's hypotheses are robust in the sense that his/her knowledge is richly interconnected and this interconnection forms the basis of his/her ability to maintain many of the

links and conclusions established during the process of context
formation. Hence, the expert can switch from considering one
hypothesis to another by a lateral process rather than "backtracking" through a tangle of hypotheses.

When an initial context has been established by the triggers,
other findings take on a new significance; **in that context**, they
act as triggers and thus contribute to the process of refining the
hypotheses of that context. For example, a headache may not,
initially, be a trigger but in the context of acute glomerulonephritis, the headache finding will trigger consideration of
hypertensive encephalopathy, which is a complication of acute
glomerulonephritis.

Computer model

Hypotheses are interconnected disorder frames of medical
knowledge. Some disorder frames represent what we might term
intermediary hypotheses of physiological states or clinical states
which, when either established or excluded, help to establish or
exclude disease hypotheses. Disease hypotheses are related to
these intermediary hypotheses, and to each other, by pointers.
Each frame has both a list of findings which might be found in a
typical example of that disorder and a list of findings which
trigger the active consideration of that disorder. Each frame has
categorical rules for both establishing and excluding the
hypothesis. Some of the frames have pointers to other frames
which are complications of the disorder which require evaluation.
There may be other pointers which represent the lines along which
the physician makes lateral switches from a hypothesis to other
hypotheses and the frame contains the conditions under which such
lateral swithes are made. Each frame contains matching scores for
the estimation of the degree to which the values of the actual
findings match the expected values of the typical findings and
thus estimate that the disorder is present in the patient to a
certain degree.

Initially, the system is only responsive to trigger findings.
When a patient-specific finding is a trigger of a hypothesis then
the hypothesis is activated. When a hypothesis is activated it
"semi-activates" related hypotheses (e.g. complications). A semi-
activated hypothesis is activated by **any** finding that is typical
of it, i.e. it could be activated by a finding that would not
otherwise act as a trigger. The active hypotheses become confirmed or inactivated by the rules for establishing or excluding
them. These rules are either the categorical rules of the frames
or rules based on an estimate that the disorder is present in the
patient. A changed pattern of activated hypotheses represents the
new context for the investigation of the patient's complaints.

Typical findings

TRIGGERS
OTHERS

Confirmation & exclusion categorical rules

IS-SUFFICIENT: <findings>
MUST-NOT-HAVE: <findings>
MUST-HAVE: <findings>

Complementary relations to other disorders

CAUSED-BY: <disorders>
CAUSE-OF: <disorders>
COMPLICATED-BY: <disorders>
COMPLICATION-OF: <disorders>
ASSOCIATED-WITH: <disorders>

Competing relations to other disorders

DIFFERENTIAL DIAGNOSIS (laterally competitive hypotheses):

$(<condition_1>,<disorders>)$ .. $(<condition_n>,<disorders>))$

Likelihood estimator

SCORES:

clause 1: $(<condition_1>,<score>)$ .. $(<condition_{n_1}>,<score>)$

clause 2: $(<condition_1>,<score>)$ .. $(<condition_{n_2}>,<score>)$
  .
  .
clause m: $(<condition_1>,<score>)$ .. $(<condition_{n_m}>,<score>)$

Fig. 4.1: Structure of a disorder frame
(adapted from Szolovits and Pauker (1979)).

4.1 STATICS

PIP's knowledge-base contains information on **findings** and **disorders**. Findings are facts about the patient; observations from physical examinations, history, results of laboratory tests, etc. Disorders (diseases, clinical states, physiological states) are not directly observable entities and need to be inferred from the reported findings. A PIP **hypothesis** represents the program's conjecture that a particular disorder is manifest with a temporal aspect (PIP's model admits only five temporal aspects: PAST, RECENT-PAST, NOW, NEAR-FUTURE AND FUTURE).

4.1.1 Disorders

Disorders are represented by frames whose structure is given in figure 4.1. (A hypothesis is an **instantiation** of a disorder frame, with the suspected time of the disorder). The slots of a PIP disorder frame fall into five categories:

Typical findings: These are the range of findings that are to be expected in a patient suffering from the disorder. Certain findings (**triggers**) are specified as strongly suggestive of the presence of the disorder.

The logical decision criteria (exclusion and confirmation rules): These are rules which enable PIP to definitely conclude the presence or absence of the disorder based on certain key findings. Logical combinations of findings (by NOT, AND and OR) may also be used to specify complex criteria.

The complementary relations: These capture various associations between the disorder and other disorders. The complementary relations are used to determine the **focus of attention** and to act as a channel for propagating likelihood scores. The disorder frames are linked into a complex network through their complementary relations.

The competing relations (differential diagnosis links): These are indications that some finding in the context of considering this disorder should suggest some other specific hypothesis instead. Through these links the disorder is associated with others that it may be confused with because they manifest similar findings, and thus constitute alternative explanations.

The numerical likelihood estimator: The scoring procedure uses the (matching) scores of the clauses to estimate the degree to which reported findings are consistent with the expected findings of the disorder.

Table 4.1: Finding Subject: Edema.

DESCRIPTOR	VALUE SET	TYPE
Status	Present, Absent	yes-no
Location	Pedal, Facial, Peri-Orbital Local, Arm, Finger, Abdomen, Generalized	multi-valued
Severity	1+, 2+, 3+, 4+, Massive	single-valued
Duration-of-Episode	Days, Weeks, Months, Years	single-valued
Pitting	Pitting, Non-Pitting	yes-no
Pain	Painful, Non-Painful	yes-no
Erythema	Erythematous, Not-Erythematous	yes-no
Symmetry	Symmetrical, Assymmetrical	yes-no
Daily-Temporal-Pattern	Without-Daily-Temporal-Pattern, Worse-in-Morning, Worse-in-Evening	single-valued
Recurrence	First-Time, Infrequent, Occasional, Frequent	single-valued
Duration-of-Pattern	Days, Weeks, Months, Years	single-valued

"Status" is a special descriptor which is associated with every finding subject. Any reported finding should at least specify its value as present or absent. It is like a flag indicating whether the values for the remaining descriptors should be sought. The most general finding of edema, "edema present", initiates a procedure for acquiring the values of the remaining descriptors. A specific finding, say, "A 2+ symmetric facial edema lasting a few days which appears worse in the mornings, has not occurred before, is non-pitting, painless, and non-erythematous", could result form this information acquisition procedure.

The knowledge-base contains over 70 frames related to some 20 different diseases and to a variety of clinical and physiological states that are associated with these diseases. Frames typically contain five to ten findings, three or four exclusionary rules, ten to twenty scoring parameters and five to ten links to other frames.

4.1.2 Findings

The set of findings is partitioned into a set of about 160 **finding subjects** represented in terms of procedures. Each finding subject is associated with a procedure that organises the information that characterizes it, viz., a set of descriptors each one of which is, borrowing MYCIN's terminology, either "single-valued" (has a set of mutually exclusive possible values), "multi-valued" (has a set of possible values that need not be mutually exclusive) or "yes-no" (the descriptor is either true or false). A finding is an association of descriptor values with descriptors. Table 4.1 specifies the finding subject **edema**; it gives the descriptors, their types and their associated value sets.

A reported finding matches a typical finding of a disorder frame if the discriptor/discriptor-value patterns are identical. A reported finding subsumes a typical finding of a disorder frame if a subpart of it matches the typical finding. A finding figuring in some disorder frame evaluates as true if either it matches with, or it is subsumed by, a reported finding. In such cases it is simply said that a prototypical finding matches a reported finding. A reported finding partially matches a typical finding if its descriptors are the same but its descriptor values are different.

A frame's score clauses contain scores for both matching and partially matching typical findings, and these are used to evaluate the fit of reported findings to findings expected on the hypothesis.

Discrimination network

In order for the system to retrieve efficiently, for any given finding, all the frame slots where that finding figures the system uses a discrimination network which associates the domain of possible findings to the domain of disorder frames. Figure 4.2 illustrates one conceptualization of this relationship.

The use of the discrimination network, demonstated in the next section, permits data-driven reasoning whilst the frame network permits hypothesis-driven reasoning.

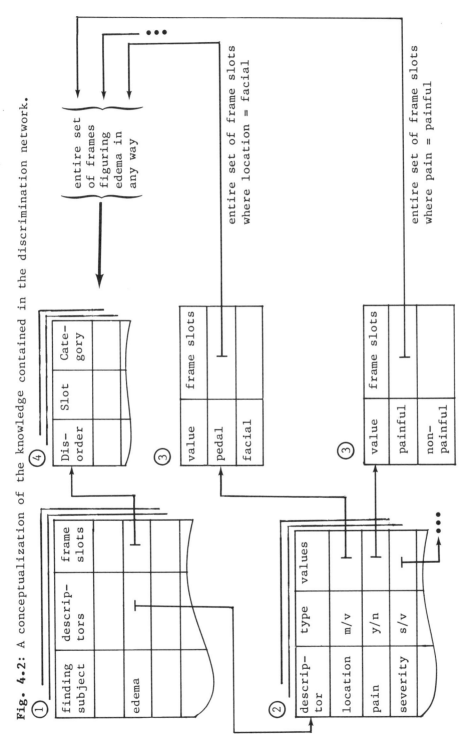

Fig. 4.2: A conceptualization of the knowledge contained in the discrimination network.

KEY: Numbers denote order of file accessing.

4.1. 3 Commonsense Knowledge

The PIP knowledge-base contains a set of rather specific rules that capture some of the physician's common sense: if the question of past proteinuria is raised, PIP can conclude its absence if the patient passed a military physical examination at that time. Such inferences are purely categorical.

4.2 DYNAMICS

An abstract description of the program's dynamics may be gained by considering the possible status transitions of hypotheses during a consultation. Figure 4.3 gives a basic hypothesis status transition diagram. Initially all disorder frames are classified as **inactive.** An inactive disorder frame is **activated** when a reported finding matches one of.its triggers. If a hypothesis is activated, all of its closely related complementary hypotheses are **semi-activated.** Furthermore, if a differential diagnosis condition is met the relevant competitor is semi-activated. A semi-active hypothesis is activated when **any** of its associated findings, and not just its triggers, matches a reported finding. Active hypotheses are assigned **matching scores** estimating the "fit" of the reported findings to the expectations (typical findings) of the hypotheses. If the matching score of some active hypothesis exceeds a preset threshold value then the hypothesis is **confirmed.** If on the other hand, the matching score is less than another preset threshold value, then the hypothesis is inactivated. Finally, a hypothesis may change its status from active to inactive/confirmed when an exclusion/confirmation criterion is met.

Thus active hypotheses are instantiations of frame disorders and constitute likely explanations for reported findings, whilst semi-active hypotheses constitute potential explanations.

4.2.1 Outlining the Overall Diagnostic Strategy

At the beginning of a consultation with PIP the user presents the system with the patient's age, sex and chief complaint/s. Once this initial stage has been completed (i.e. once the chief complaint/s has/have been fully characterized), the system cycles through a data-driven reasoning mode followed by a hypothesis-driven reasoning mode. This cycle can be interrupted. The user may enter unsolicited input; if so the cycle restarts as soon as the information acquisition task (dealing with the new input) has been completed. The consultation terminates when all the active

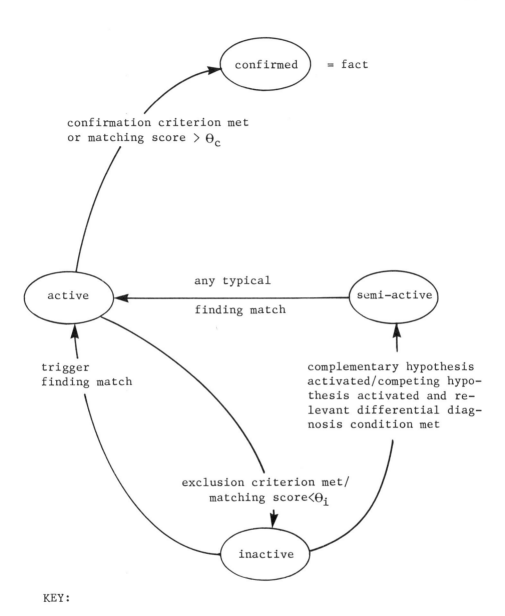

KEY:

Θ_c/Θ_i: threshold value for confirmation/inactivation

Fig. 4.3: Basic hypothesis status transition diagram.

and semi-active hypotheses have been fully explored. This termination criterion leads to many irrelevant questions being asked.

Data-driven reasoning

In this mode of reasoning PIP is processing a newly acquired finding by accessing the discrimination network. The finding is processed as discussed below (the reader may find it useful to cross refer to figure 4.2). By the end of this processing the configuration of the hypothesis space could be altered and perhaps the system's focus of attention is shifted to a different hypothesis.

The finding subject of the finding is identified and the discrimination network is used to identify all the frame slots figuring findings of that finding subject. The next step determines which of these findings match with the newly acquired finding. This is done by using the values assigned to the descriptors to eliminate frame slots that list findings involving different values for the descriptors than the reported values. For example, if the reported value for a "yes-no" descriptor is "yes" all the frame slots that list findings involving a "no" value for that descriptor are eliminated.

The matched findings are used to **revise** the configuration of the hypothesis space according to the transition conditions specified in figure 4.3. For example, if such findings figure under the numerical likelihood estimator of some active hypothesis, then this causes the revision of the hypothesis' likelihood estimate which, in turn, may cause the hypothesis' inactivation or confirmation.

Hypothesis-driven reasoning

When in this mode of reasoning, PIP is exploring the most likely active hypothesis as explained below; this results in acquiring, for processing, a new finding about the patient. In this mode the system is utilising the network which is implicitly defined by the disorder frames.

4.2.2   Focusing and Information Acquisition Strategies

PIP requires that a hypothesis be activated on the basis of either only strong direct evidence (triggers) or by a strong combination of circumstantial and (possibly not, in itself, very strong) direct evidence. This is how the system constrains its scope of attention. The most promising active hypothesis and its

associated complementary hypotheses form the focus of attention,
i.e. the most promising active hypothesis is used as a basis for
eliciting further observations from the user (hypothesis-driven
reasoning). The promise of an active hypothesis is expressed as
an updateable numerical value which is the current estimate of the
likelihood of the hypothesis being true for the patient. The most
promising hypothesis is explored by eliciting information concerning one of its, as yet, unexplored typical findings. If all its
findings have been explored, the system attempts to elicit
circumstantial evidence by exploring some finding of the highest
ranking hypothesis which is complementary to the leading
hypothesis. The differential diagnosis links constitute a mechanism for shifting the focus of attention.

PIP's information acquisition strategy requires that on
acquisition of a new finding from the user, its finding subject be
identified and the user be further queried for values of the
entire set of descriptors associated with the finding subject
(i.e. the procedure that characterizes the finding subject is
invoked). This strategy is followed when the system is exploring
typical findings of active hypotheses, i.e. the system not only
elicits values for the relevant descriptors, but for the entire
set of descriptors associated with the relevant finding subject.

Below we discuss PIP's scoring procedure.

Estimating the likelihood of active hypotheses

The **likelihood estimate** of some active hypothesis is a
combination of a measure of the "explanatory power" of the
hypothesis, vis-a-vis all the observed findings (**binding score**)
and a measure of the confidence that the observed findings fit the
expectations of the hypothesis (**matching score**). Thus the most
likely hypothesis is the one which explains the widest range of
strongly matched reported findings.

The function that gives the likelihood estimate, L_H, of an
active hypothesis H, is the average:

$$L_H = (B_H + M_H)/2$$

where M_H and B_H are respectively, the matching and binding scores
of H.

In PIP it is not necessary to list every expected finding of
some disease within the frame of that disease. Clusters of
findings that tend to co-occur in instances of many different
diseases are made the typical findings of a clinical state which
is associated with the relevant diseases. However, this compli-

cates the computations for the matching and binding scores. To "undo" the organisation imposed by the use of clinical and physiological states, when active, the scores of these states are "pumped" into the relevant active disease hypotheses.

Binding score

The wider the range of reported findings that are expected on a hypothesis the wider is that hypothesis' explanatory power.

The exlanatory power is measured by the binding score:

$$B_H = \frac{\text{number of reported finding expected on H}}{\text{total number of reported findings}}$$

A reported finding is expected on H if either it is directly expected on H (forms a typical finding of H) or it is expected on one of H's closely related active hypotheses. The binding score of hypothesis H is thus computed in two stages; first the "local binding score" is computed and then the binding scores from related clinical and physiological states are pumped into H.

Local binding score

The local binding score Lb_H of hypothesis H is given by:

$$Lb_H = \frac{\text{number of reported findings directly expected on H}}{\text{total number of reported findings}}$$

Propagating binding scores

Assuming that the binding scores of clinical states are given by their local binding scores, the binding score for H is given by:

$$B_H = Lb_H + \sum_j B_{H_j}$$

where: j ranges over the set of active clinical and physiological states related to H.

Matching score

Like the binding score, the matching score of hypothesis H is computed in two stages; first the "local matching score" is taken and then the matching scores from related clinical and physiological states are pumped into H.

Local matching score

The local matching score is computed using the hypothesis'
numerical likelihood estimator. Numerical likelihood estimators
are attached to a series of clauses; below we show how:

Taking the case of a finding **descriptor** which is "single-
valued" (or "yes-no") (e.g. AGE, SEX, etc), evidence for a
particular descriptor value is evidence against the other values
being realized. For example, a patient is either a child, middle
aged, or old but not both child and old. Thus when findings are
associated with a particular frame, the descriptor values may be
ordered as to their fit with the frame's expectation. For
example, if a frame expects the value of the descriptor AGE to be
CHILD, then the values of MIDDLE-AGED and OLD will be
progressively counter-indicative of the frame hypothesis. The
evidential fit of the actual value to the frame's expectation is
specified in a clause with evidential weights attached to each
possible value. For example, the numerical likelihood estimator
for Acute Glomerulonephritis contains the following clause:

```
(PATIENT with AGE = CHILD or YOUNG →    0.8)
(PATIENT with AGE = MIDDLE-AGED     →  -0.5)
(PATIENT with AGE = OLD             →  -1.0)
```

Hence the patient being OLD is strong evidence against
(indicated by a "minus sign") the disorder being Acute Glomerulonephritis.

In other cases, where a finding has more than one descriptor,
evidential weight is attached to the set of descriptors. Where
the reported finding subsumes the expected finding of a hypothesis
it is taken to match the typical finding, and provides at least as
much evidential support for the hypothesis. Where the expected
finding subsumes the reported finding then the match is not as
great and, therefore, the evidential support is lessened. For
example, the numerical likelihood estimator for Acute Glomerulonephritis contains the following clause:

```
(EDEMA with LOCATION = FACIAL or PERI-ORBITAL,
            SYMMETRY = not ASYMMETRICAL,
            PAINFULNESS = not PAINFUL,
            ERYTHEMA = not ERYTHEMATOUS,
            DAILY-TEMPORAL-PATTERN = WORSE-IN-MORNING) → 1.0

(EDEMA with LOCATION = FACIAL or PERI-ORBITAL,
            SYMMETRY = not ASSYMETRICAL,
            PAINFULNESS = not PAINFUL,
            ERYTHEMA = not ERYTHEMATOUS) → 0.5
```

In this case, the observation of the second finding is sufficient to trigger the hypothesis of Acute Glomerulonephritis (i.e. the finding is a typical trigger finding for the disorder). However, the additional observation that "edema is worse in the morning" increases the strength of this evidential support by a factor of two.

Figure 4.4 gives a pictorial representation of the contents of the numerical likelihood estimator of some hypothesis, H.

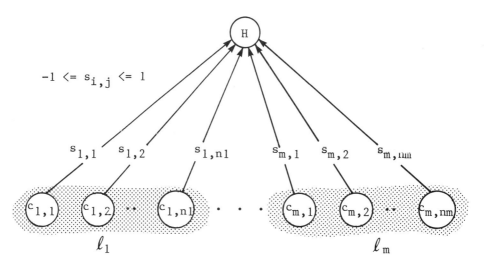

KEY:

ℓ_i; $i=1,\ldots,m$: the ith clause.
$(c_{i,j}, s_{i,j})$; $i=1,\ldots,m$; $j=1,\ldots,n_i$: the jth condition of the ith clause and its associated evidential weight or score $(s_{i,1} > s_{i,2} > \ldots > s_{i,ni})$. It can be seen that a clause is a sequence of conditions and evidential weight associations **sorted in descending order.**

Fig. 4.4: Contents of numerical likelihood estimator.

The support of the relevant findings to a hypothesis is calculated from the evidential weights found in its clauses. The value/s of the finding's descriptor (or descriptor set) selects the evidential weight and thus the contribution that a clause makes to the overall support.

The computation of the overall (local) support takes the advantage of the fact that conditions within a clause are sorted in descending order of evidential weights. Thus the local matching score, Lm_H, is given by the ratio of the actual to the maximum possible score:

$$Lm_H = \left\{ \sum_{i=1}^{m} S_i \right\} / \sigma \; ; \; -\infty < Lm_H <= 1$$

where: $\sigma = \sum_{i=1}^{m} s_{i,1}$ is the maximum possible total score, and

S_i is the contribution of clause ℓ_i given by:

$$S_i = \begin{cases} s_{i,k} & \text{-- if } c_{i,k} \text{ is the first condition in the} \\ & \text{sequence of conditions } \ell_i \text{ that evaluates to} \\ & \text{true (i.e. matches a reported finding).} \\ \\ 0 & \text{-- if none of } \ell_i\text{'s conditions} \\ & \text{evaluates to true.} \end{cases}$$

Propagating matching scores

The active/confirmed clinical and physiological states contribute a matching score to all those active hypotheses in which they figure. Consider the simple abstract case in figure 4.5. Assuming that the matching scores of clinical states are given by their local matching scores, then the matching score for H is given by:

$$M_H = \frac{a + a_1 + a_2 + a_3}{\sigma + \sigma_1 + \sigma_2 + \sigma_3}$$

which is precisely the undoing of the organisation imposed by the clinical and physiological states. Note that $M_{H_3} = (a_2 + a_3)/(\sigma_2 + \sigma_3)$ is a component of M_H.

The above model of score propagation could be complicated by having associational strengths defined on the links. This, however, would be at least, inelegant because it would mix direct evidence provided by findings with circumstantial causal evidence.

As PIP's information acquisition strategy is based solely on the focusing likelihood scores, and as the top scoring hypothesis tends to change frequently, the program's questioning can be arbitrary. A more principled information acquisition strategy,

which asked a carefully planned cluster of questions about the
current focus **before** re-evaluating the hypothesis space, would be
more satisfactory.

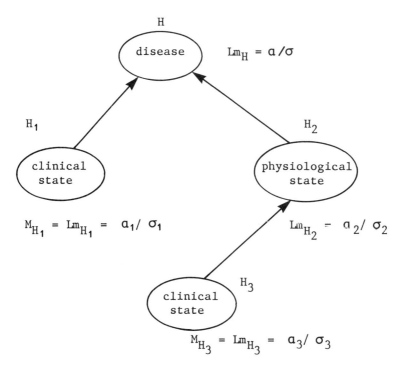

Fig. 4.5: Active/confirmed hypotheses related to H.

4.3 IMPLEMENTATION DETAILS

The program was originally written in the CONNIVER language,
a superset of MACLISP. It ran on a DEC PDP-10 computer. Since
that time, the program has been partially re-written directly in
MACLISP and is now much more efficient. The original version
suffered from the problems inherent to a distributed control
structure. The revised program uses a central control structure
and uses a discrimination network to accomplish pattern matching.
In the initial version of the program, the problem of pattern
matching was handled by the CONNIVER data base, which was not
designed for this complex pattern matching task and which was
consequently unacceptably slow. The use of the discrimination
network in the current version of the program gave rise to a
considerable speed increase in the execusion of this task.

REFERENCES

Elstein A.S., Shulman L.A. and **Sprafka S.A.** (1978): Medical
Problem Solving: an analysis of clinical reasoning, Harvard
University Press, Mass.

Pauker S.G. Gorry G.A. Kassirer J.P. and **Schwartz W.B.** (1976):
"Toward the simulation of clinical cognition: taking a present
illness by computer", **The American Journal of Medicine 60,** pp.
981-995.

Pauker S.G. and **Szolovits P.** (1977): "Analyzing and simulating
taking the history of the present illness: context formation", in
Schneider/Sagvall Hein (eds.), **Computational Linguistics in
Medicine,** North Holland, pp. 109-118.

Szolovits P. and **Pauker S.G.** (1978): "Research on a medical
consultation system for taking the present illness", Proc. of the
Third Illinois Conference on Medical Information Systems, University of Illinois at Chicago Circle, pp. 299-320.

Szolovits P. and **Pauker S.G.** (1978): "Categorical and probabilistic reasoning in medical diagnosis", **Artificial Intelligence,**
Vol. 11, pp. 115-144.

Chapter 5
INTERNIST-1

Application area:	Medicine.
Principal researchers:	J.D. Myers and H.E. Pople (University of Pittsburgh).
Function:	Diagnosis of internal medicine.

OVERVIEW

The clinical problem

The physician needs expert assistance where he/she might experience difficulty in formulating the diagnostic problem as an analytic task of dealing with a fixed set of diagnostic alternatives (differential diagnosis).

The work of the INTERNIST/CADUCEUS (Pople,1982) project has concentrated on the investigation of heuristic methods for imposing differential diagnostic task structure on the clinical decision making problem. An aspect of this investigation is to devise knowledge structures that aggregate groups of diseases/ pathological states into a taxonomy. This taxonomy may then be used to select or rule out groups of diseases on the basis of an answer to a question directed at a common characteristic. In INTERNIST-I (Pople, 1975 and Miller **et al**, 1982) the taxonomy is a strict hierarchy, organised around the concept of organ system involvement. It has been found that a strict hierarchy can not be used effectively to develop foci during the course of a consultation. Also the approach of using only a taxonomy of diseases is deficient, relative to the precision found in a causal model, in assessing the attribution of findings to particular diseases. Current work on CADUCEUS (see below) is aimed at investigating the conceptualisation of the differential diagnostic problem in both causal and non-strict hierarchical terms.

The INTERNIST-I knowledge-base

The knowledge is of two basic types: diseases and manifestations (e.g. history, symptoms, physical signs and laboratory findings). The knowledge-base has been structured so that abnormal findings are used to abduce disease hypotheses directly without calling for consideration of any intermediate pathological conditions. The manifestations of the various pathological states associated with each disease, are incorporated within the disease's profile. The exceptions are various well established states of clinical significance (e.g. heart failure) which figure in framing a differential diagnosis.

Each manifestation defines a set of diseases by means of an EVOKES data structure. This structure records, for each manifestation the list of diseases in which the manifestation has some likelihood of occurring. The strength (0 to 5) by which the manifestation is linked to the disease is contained in the data structure. Another data structure, MANIFEST, profiles each disease with a list of manifestations of the disease and an estimate (1 to 5) of their frequency of occurrence.

Other relations are defined on the set of diseases. These contain the causal, temporal and other associations by which various diseases and pathological states are interrelated. These relations also incorporate weighing factors similar to those of EVOKES and MANIFEST.

An important auxiliary relation, IMPORT, expresses (1 to 5) the global importance of a manifestation, i.e. the importance attached to the fact that the manifestation be explained by a concluded hypothesis.

The problem-formation method

Initially patient data are entered. The positive findings define a set of disease/disease categories. Where a manifestation evokes one of the disease categories it is because that finding can be explained on the hypothesis of any subdivision of the category.

In order to assess the attribution of a finding to a particular disease, the program uses a scoring procedure that awards "credit" to a disease on the basis of its "explanatory power". Thus the evocative strength and global importance of manifestations explained by it are counted in its favour, whereas the frequency of occurrence weights, of those manifestations expected but not found present, count against.

From the ordered list of disease hypotheses a new task can be constructed. The highest order hypotheses can be considered alternatives if they can individually explain as many findings as

they could taken together. This set of alternatives becomes the
differential diagnosis which is the focus of the problem solving
activity. If there is only one element in the set of highly
ranked alternatives then the program pursues this hypothesis
seeking confirmatory data to further separate it from the others.
When this "distance" is sufficiently large or there are no more
useful questions to be asked, the hypothesis is confirmed. In the
case where there are a number of highly ranked alternatives the
program looks for data that will discriminate between them and
hence widen the distances between them. Where the number is
greater than five the program first seeks negative findings to
rule out one or more elements. If, when there is more than one
element in the set of alternatives, the system runs out of useful
questions, then the hypotheses are deferred, and the set of
findings associated with them marked as likely to be explained by
some unknown element of that differential diagnosis. In solving a
second, perhaps related, problem, the program may be able to
return to the deferred problem and make more progress.

Concluded hypotheses are entered on a list, and manifestations explained by them marked as no longer in need of
explanation.

Because of causal, temporal and other interrelationships,
certain disease combinations are more likely to occur than others;
hence on confirmation of a disease hypothesis linked hypotheses
are given additional weight and thus may require consideration.

The cycle is repeated until all problems present in the case
have been revealed and their solution found.

5.1 STATICS

INTERNIST-I has a hierarchically organized knowledge-base
which is built around the concept of a disease profile.
Diseases/disease categories (and important high level pathological
and clinical states) are profiled in terms of clusters of
manifestations. The word 'manifestation', in the context of
INTERNIST-I, is ambiguously used to denote either a finding
(historical item, symptom, physical sign, laboratory abnormality),
a predisposing factor, or a disease related to a disease process.
Disease profiles are captured within frame-like structures, the
format of which is given in figure 5.1.

The basic conceptual structure of the system's knowledge-base
is given in terms of the following relationships (depicted in
figure 5.2):

SPECIALISATIONS

FORM-OF: <diseases>

MANIFESTATIONS

findings:

(<finding$_1$><evocative value><frequency value>)...
(<finding$_n$><evocative value><frequency value>)

complementary relations to other diseases:

PREDISPOSES-TO:
    (<disease$_1$><evocative value><frequency value>)...
    (<disease$_n$><evocative value><frequency value>)

CAUSES:
    (<disease$_1$><evocative value><frequency value>)...
    (<disease$_n$><evocative value><frequency value>)

COINCIDENT-WITH:
    (<disease$_1$><evocative value><frequency value>)...
    (<disease$_n$><evocative value><frequency value>)

PRECEDES:
    (<disease$_1$><evocative value><frequency value>)...
    (<disease$_n$><evocative value><frequency value>)

Fig. 5.1: Format of INTERNIST-I disease "frame"
(adapted from Miller **et al**, 1982).

EVOKES relates the domain of manifestations to the domain of
diseases/disease categories. A manifestation M is related to some
disease D under EVOKES to a specified degree denoting the
likelihood that the observation of manifestation M is indicative
of disease D. This degree, referred to as M's **evocative
association** (strength) for D, is represented as an integer from
{0,1,2...,5}. These strengths of association tend to be
subjective estimates on the basis of extensive clinical
experience. An evocative association of 5 represents the case
that the manifestation is sufficient to warrant the conclusion
that disease D is present. An evocative association of 0

represents the case that the manifestation provides no indication
that D is present.

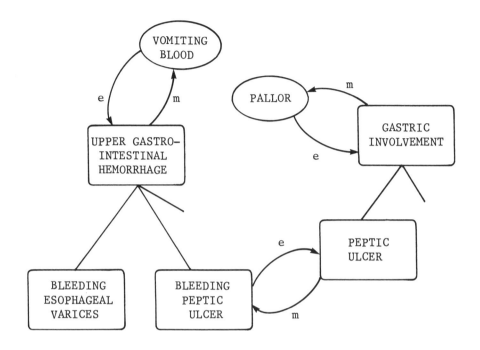

KEY:
    ──── : FORM-OF; ──e►: EVOKES; ──m►: MANIFEST;
    ◯ : finding; ▭ : disease/disease category.

MANIFEST and EVOKES relations can exist between diseases;
this reflects the fact that one disease may be caused-by, and
hence be a manifestation of another disease.

Fig. 5.2: Illustrating the conceptual structure of INTERNIST-I's
        factual knowledge (adapted from Pople, 1982).

The list of diseases related to a manifestation M, under EVOKES,
forms the **"differential diagnosis list"** of M. EVOKES is used for
data-driven reasoning -- from manifestations hypotheses may be
abduced.

    MANIFEST relates the domain of diseases/disease categories to
the domain of manifestations. A disease D is related to some
manifestation M under MANIFEST to a specified degree denoting how
often M is associated with instances of M. This degree of

association is referred to as the **frequency of occurrence** of M in
D and is given as an integer from $\{1,2,..,5\}$. Such estimates can
be supported by data obtained from a review of the relevant
medical literature and in this sense can be more objectively
based. A frequency of occurrence of 5 represents the case that
the manifestation is necessary in the disease (i.e. if not M then
not D). A frequency of occurrence of 1 represents the case that
the manifestation occurs rarely in the disease (hence not M hardly
weighs as a counterindication of D). The list of manifestations
related to a disease D under MANIFEST represents D's **profile.**
MANIFEST is used for hypothesis-driven reasoning; through a
disease profile we may deduce the manifestations which are likely
to occur on the disease being present in the patient.

FORM-OF relates disease categories to their underlying
specialisations. FORM-OF(D1,D1.1) represents the fact that D1.1
is a form (specialisation) of disease category D1. The manifestations for a disease category are those common to each of its
specialisations (for the disease category the evocative association and frequency of occurrence of each manifestation, is,
respectively, the maximum evocative association and minimum
frequency of occurrence of that manifestation among the specialisations). The disease taxonomy constructed by the FORM-OF
relation is on the basis of the concept of organ system
involvement.

IMPORT is a disease independent attribute associated with
manifestations. The IMPORT, expressed as an integer from
$\{1,2,..,5\}$, of a manifestation denotes the global importance of
the manifestation, and thus the extent to which one is compelled
to explain its presence in any patient. An IMPORT value of 5
represents the case that the manifestation absolutely must be
explained by one of the concluded diseases. An IMPORT value of 1
represents the case that the manifestation is usually unimportant,
for example, it occurs commonly in normal persons and thus can
easily be disregarded.

Knowledge of how the presence or absence of a manifestation
may influence the presence or absence of other manifestations is
also included in the knowledge-base.

The domain of internal medicine is large and diverse.
Presently the system's knowledge-base, which represents 15 person-
years of work, covers 80% of the diseases of internal medicine,
making INTERNIST-I the largest Artificial Intelligence in Medicine
(AIM) system. More specifically the knowledge-base encompasses
about 600 individual disease profiles and approximately 3550
manifestations. There are about 2600 links between diseases and
6500 links between manifestations.

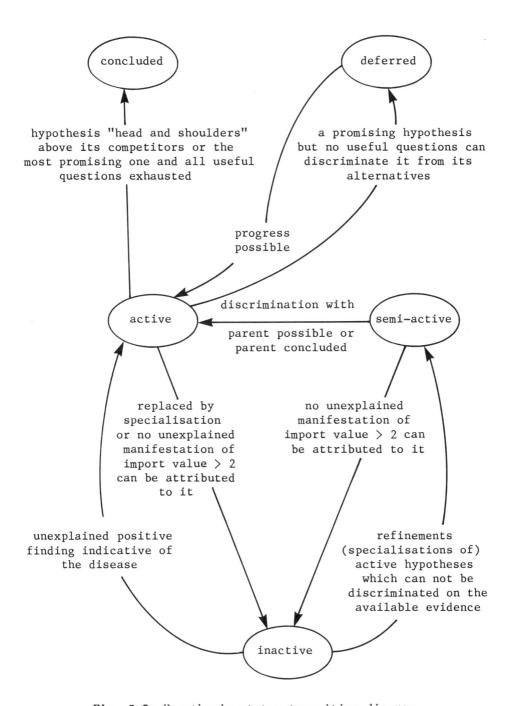

Fig. 5.3: Hypothesis status transition diagram.

5.2 DYNAMICS

The system employs abductive (data-driven) and deductive (hypothesis-driven) reasoning in order to establish the presence of particular diseases in the patient. Manifestations constitute the evidence and diseases/disease categories the hypotheses.

The dynamics of the system can be abstracted in terms of the hypothesis status transition diagram of figure 5.3. Initially all hypotheses are **inactive.** A hypothesis is **activated (evoked)** when at least one unexplained finding which is indicative of the disease, under EVOKES, is observed to be present in the patient. Disease categories are active when no information to enable the discrimination between their underlying specialisations is yet available. Specialisations of an active disease category can be considered to be **semi-active** hypotheses waiting to replace their category in the active status as soon as evidence supporting them becomes available. Thus the disease taxonomy is an important mechanism for controlling the proliferation of active hypotheses during the diagnostic process. A sufficiently promising active hypothesis is **concluded** either if it has no competitors (where the competitors of a disease/disease category are not predefined but are dynamically determined) or if it is sufficiently more promising than its nearest competitor. The conclusion of a hypothesis causes all observed manifestations explained by it to be removed from future consideration; subsequent observations are marked as explained when a previously concluded hypothesis can account for them. Concluding a disease category means that the observed manifestations explained by it can only be subsequently attributed to specialisations of it -- intuitively the conclusion of a disease category would cause the activation of its underlying semi-active specialisations. The most promising active hypothesis and its competitors are transferred into the **deferred** status when the level of questioning, permitted for the given set of alternatives (see section 5.2.3), has been exhausted, and the set's resolution has not been possible; the set of observed manifestations associated with the given set of hypotheses are marked as likely to be explained by some unknown element of that group of alternatives. A deferred set of hypotheses is returned to the active status when subsequent investigation with other active hypotheses enables the progression of the set's resolution process. Finally an active/semi-active hypothesis is inactivated when no unexplained manifestation with a global import value of at least 3 can be accounted by the hypothesis.

The dynamics of the system suffer from the following shortcomings:

The system is not in a position to perceive the multiplicity of diseases in a case at once; problem areas (see section 5.2.1) are constructed in a sequential rather than parallel fashion. Only after a specific disease is concluded, can bonus marks be assigned to causally related diseases in separate problem areas (see section 5.2.2).

When a disease is concluded, that disease is allowed to explain any observed manifestations that are listed on its profile. Once explained, a manifestation is no longer used to evoke new disease hypotheses or to participate in the scoring process. INTERNIST-I, therefore, assumes (quite wrongly) that in any patient any manifestation can have only one cause. Also it is not possible to undo a previous diagnostic conclusion when contradictory evidence becomes available.

The diagnostic accuracy of the system is weakened (constrained) by the representation of diseases as exhaustive lists of manifestations. If diseases were represented in terms of their intermediate states (disease evolution in time), the presence or absence of each state would be independently determined, and a disease would be allowed to explain a manifestation only when a state causing the manifestation was confirmed. Furthermore, as a predisposing factor is listed as a manifestation in the disease profile, the disease can "explain" the predisposing factor, when, should the predisposing factor be observed, a search for an independent cause should be made.

Thus, the dynamics of the system could be improved by improving the structure of its factual knowledge (knowledge-base). The design of the intended successor of INTERNIST-I, CADUCEUS, is an attempt to improve the quality of INTERNIST-I's diagnostic reasoning by building a more highly structured knowledge-base that would permit such improved reasoning activity.

5.2.1 Outlining the Overall Diagnostic Strategy

At the beginning of a consultation with INTERNIST-I, the user enters all **positive** (observed to be present) and **negative** (observed to be absent) findings considered significant. Findings are expressed as precise sequences of terms in a controlled vocabulary. Next, the system alternates between data- (abductive) and hypothesis- (deductive) driven reasoning until every positive finding with a global import value of at least 3 has been attributed to one of its potential causes. Thus every positive finding instigates an **elementary diagnostic task,** for the attribution of the finding to a disease that explains it.

Data—driven reasoning

In this mode of reasoning the system is abducing hypotheses from manifestations whose presence in the patient has not yet been explained. The differential diagnosis list for every positive manifestation is retrieved from the knowledge-base and hypotheses are activated/semi-activated according to the conditions specified in figure 5.3. Negative manifestations are not used for abducing hypotheses, but they are used in evaluating the promise of hypotheses abduced from the positive manifestations (see section 5.2.2).

Hypothesis—driven reasoning

In this mode of reasoning the system is basically seeking to confirm the deductions that are drawable from active promising hypotheses, i.e. attempts to elicit from the user additional information concerning manifestations expected on the hypotheses, and thus confirm hypotheses.

The promises of active hypotheses are evaluated using the unexplained manifestations and negative findings. The **competitors** of the most promising hypothesis are dynamically determined by a **partitioning** heuristic. The most promising hypothesis and its competitors form the system's current focus of attention or the current **problem area** to be resolved. The construction of a problem area is equivalent to synthesizing the relevant set of elementary diagnostic tasks into the single, higher level differential diagnostic task, of determining which alternative in the problem area is present in the patient. If the configuration of the problem area does not permit the conclusion of the most promising hypothesis, the system attempts to elicit additional useful items of information from the user. Any new manifestations thus elicited, whose presence in the patient can not be explained by an already concluded hypothesis, cause the system to switch back to its data—driven mode of reasoning.

The INTERNIST-I diagnostic strategy, therefore, involves the repeated scoring and partitioning of active hypotheses in the light of new evidence. Hence the system's focus of attention could switch from one problem area to another when questioning in the first area has been counterproductive.

5.2.2 Focusing Mechanism

The INTERNIST-I focusing mechanism determines the system's current focus of attention or problem area by: firstly scoring the active hypotheses, secondly temporarily discarding as unattractive those active hypotheses whose scores fall a threshold number of points below the most promising hypothesis, and finally partition-ing the remaining hypotheses into competitors and non-competitors of the most promising hypothesis. The most promising hypothesis and its competitors form the current problem area. Figure 5.4 shows the various subcategories to which active hypotheses are allocated by the system's focusing mechanism.

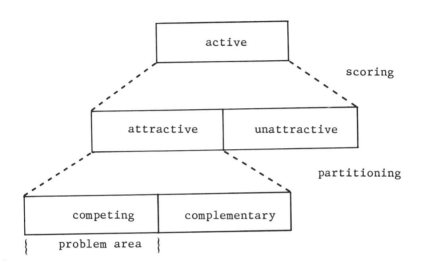

Fig. 5.4: Subcategories of active hypotheses.

Scoring of active hypotheses

Four lists are maintained for each active hypothesis (see figure 5.5):

l_1 — list of (unexplained) positive findings matching the disease profile, i.e. manifestations potentially explained by the disease hypothesis;

l_2 — list of findings which are part of the disease profile but are currently missing from the patient (possibly these are findings yet to come depending on the state of the disease reached, but as mentioned earlier, no such knowledge is held in the knowledge-base);

l_3 — list of (unexplained) findings which are present in the patient but not explained by the disease hypothesis (possibly explained by a coexistent disease?);

l_4 — list of findings which are part of the disease's profile but about which nothing is known (this list is used in determining which questions to ask next).

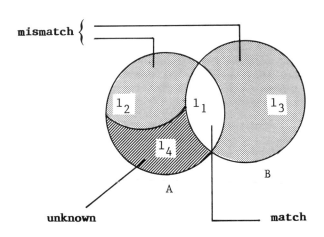

KEY:

A : profile of some disease (hypothesis);
B : positive findings (direct evidence).

Fig. 5.5: Venn diagram representation of the "explanatory power" of an active disease hypothesis.

The initial score of an active hypothesis is computed using the **match** and **mismatch** between the hypothesis' expectations and the currently unexplained positive findings. The score of the hypothesis is then completed by adding the bonuses awarded by it being the antecedent of concluded diseases. The hypothesis with the highest score is the most promising one. The details of the INTERNIST-I scoring procedure are given below:

The score of the active hypothesis H_i, $S(H_i)$ is given by:

$$S(H_i) = P(H_i) - N(H_i) + BONUS(H_i)$$

where: $P(H_i)$ is the likelihood in the presence of H_i derived from direct evidence,

$N(H_i)$ is the likelihood in the absence of H_i derived from direct evidence, and

Table 5.1

evocative association, E	$w_e(E)$ (weight of explained manifestation)
0	1
1	4
2	10
3	20
4	40
5	80

Table 5.2

frequency of occurrence, F	$w_u(F)$ (weight of missing manifestation)
1	1
2	4
3	7
4	15
5	30

Table 5.3

clinical importance, I	$w_u(I)$ (weight of unexplained manifestation)
1	2
2	6
3	10
4	20
5	40

In each case the interval, between any two consecutive points (a,b) in the scales, is $2a \simeq b$.

$\text{BONUS}(H_i)$ is the likelihood in the presence of H_i derived from circumstantial causal evidence.

$P(H_i)$ is given by:

$$P(H_i) = \sum_{j=1}^{|l_{1i}|} w_e(E_j)$$

where: j ranges over the elements of the l_1 list for hypothesis H_i, l_{1i},

E$_j$ denotes the evocative association between the jth element of l_{1i} and H_i, and

w_e is the discrete weight function analyzed in table 5.1.

$N(H_i)$ is given by:

$$N(H_i) = \sum_{j=1}^{|l_{2i}|} w_m(F_j) + \sum_{j=1}^{|l_{3i}|} w_u(I_j)$$

where: j ranges over the elements of the l_2/l_3 list for hypothesis H_i, l_{2i}/l_{3i},

F_j denotes the frequency to which the jth element of l_{2i} occurs with instances of H_i,

I_j denotes the global clinical importance value of the jth element of l_{3i}, and

w_m and w_u are discrete weight functions analyzed, respectively, in tables 5.2 and 5.3.

$\text{BONUS}(H_i)$ is given by:

$$\text{BONUS}(H_i) = \sum_{j} 20 * F_j$$

where: j ranges over the established consequences of H_i,

and

F_j denotes the frequency to which the concluded disease H_j occurs with instances of H_i.

INTERNIST-I therefore pieces together interdependent components of a multisystem illness one by one by promoting the
consideration of diseases related to previously concluded diseases
('sequential' reasoning).

The shortcomings of the INTERNIST-I scoring procedure are:

Only those manifestations which are either definitely present
or absent, can provide evidence in favour of or against some
hypothesis. This is very inappropriate when the manifestation
refers to a causally related disease, since it ignores the
arguably real support given to some hypothesis from a strongly
suspected though not confirmed consequent.

The interdependencies of manifestations are not taken into
consideration. This results in the assignment of redundant points
to diseases. If, for example, manifestations M_1 and M_2 are
observed to be present, but M_1 implies M_2 (e.g. M_1 subsumes M_2),
then only M_1 should provide evidence in favour of some disease
hypothesis whose profile contains both M_1 and M_2.

The procedure assigns credit to all manifestations, however
rare that association is.

Partitioning of attractive hypotheses

The partitioning of the attractive hypotheses is based on the
following heuristic (Miller **et al**, 1982, p.471):

"Two diseases are competitors if the items not explained by one
disease are a subset of the items not explained by the other,
otherwise they are alternatives (and may possibly coexist in the
patient). To paraphrase, if disease A and disease B, taken
together, explain no more observed manifestations than either one
does when taken alone, the diseases are classified competitors."

The leading hypothesis and its competitors constitute the
system's focus of attention for the given stage of the
consultation.

5.2.3 Information Acquisition Strategies

If the configuration of the problem area does not permit the
conclusion of the leading hypothesis, the system employs an
information acquisition strategy, the choice of which depends on
the number of competing hypotheses and their relative promises.
The following strategies are used for resolving problem areas:

The **RULEOUT** strategy is employed when there are five or more diseases within 45 points of the leading disease. This strategy restricts the questioning to findings obtainable via history or physical examination, and which are expected on the competing hypotheses to high degrees (reflected by high frequency of occurrence values). Negative responses are welcomed since they will remove (rule out) some diseases from contention.

The **DISCRIMINATE** strategy is employed if there are two to four diseases within 45 points of the leading disease. This strategy formulates questions that serve to support one hypothesis at the expense of another; more costly procedures may be called for in order to achieve this objective; i.e. the questions may refer to findings obtainable via laboratory tests.

The **PURSUE** strategy is employed if the second best contender is 46-89 points behind the leading disease. This strategy formulates questions that serve to establish the leading disease. The level of questioning is unconstrained, so that questions about biopsies (if useful) can be asked, or other specialised procedures capable of providing pathognomonic data (fingings with evocative associations of 5).

The application of an information acquisition strategy results in the formulation of a question which is subsequently put forward to the user. Two formats are used for asking questions: the user is either asked to specify whether a specific finding is present or absent, or to provide any data that might be available within a specified category of findings. In the latter case, the user is not constrained to the suggested category of findings but is free to enter whatever positive or negative findings desired.

5.3 FACILITIES

5.3.1 Teaching Facilities

The EXAMINER system (Oleson,1977) teaches the INTERNIST-I knowledge to students. It presents the student with a list of manifestations from a case in internal medicine and then elicits the student's characterization of the disease processes present in the case. Finally it presents the student with a commentary regarding the characterization.

5.4 IMPLEMENTATION DETAILS

INTERNIST-I is implemented on a PDP-10 under the TENEX operating system. It takes about 800 mbytes. The primary implementation language is INTERLISP with interfaces to the knowledge-base coded in assembly language. Program execution times for INTERNIST-I typically range from three to seven minutes of CPU time for reasonably complex case analyses.

REFERENCES

Miller R.A. Pople H.E. and **Myers J.D.**(**1982**): "INTERNIST-I: an experimental computer-based diagnostic consultant for general internal medicine", **New England J. of medicine, Vol. 307, No. 8,** pp. 468-476.

Oleson C.E. (**1977**): "EXAMINER: a system using contextual knowledge for analysis of diagnostic behaviour", Proc. **IJCAI-77,** pp. 814-818.

Pople H.E. (**1975**): "DIALOG: A model of diagnostic logic for internal medicine ", Proc. **IJCAI-75,** pp. 848-855.

Pople H.E. (**1982**): "Heuristic methods for imposing structure on ill-structured problems: the structuring of medical diagnostics", in: Szolovits P. (ed.), **Artificial Intelligence in Medicine,** AAAS Symposium Series, Boulder, Colo: Westview Press, pp. 119-185.

Chapter 6
CADUCEUS

Application area:	Medicine.
Principal researchers:	J.D. Myers and H.E. Pople (University of Pittsburgh).
Function:	Diagnosis of internal medicine.

OVERVIEW

Although not all aspects of the CADUCEUS system (Pople 1977 and 1982) have been fully implemented, the system is not mere speculation either. It is the continuation of the INTERNIST research project and is designed to remedy the perceived shortcomings of INTERNIST-I.

There are two needs to be met in the diagnostic process and these involve some element of trade off against each other. First there is a need to quickly focus on likely hypotheses and secondly there is a need to justify the attribution of findings to the hypotheses. In CADUCEUS two knowledge structures are used to satisfy these needs. There is a nosological structure which provides a good general characterization of the clinical problem, and there is a causal network which enables the critical evaluation and justification of hypotheses.

The nosology is not the categories of diseases in INTERNIST-I but is a structure of categories of involvement where any given disease may be classified in as many descriptive categories as is appropriate.

CADUCEUS exploits data that is indicative of either pathological or nosological states (for example, the observation of jaundice is indicative of liver involvement), and sets up a number of parallel differential diagnostic **analyses**. Elementary tasks involve deciding between the subclassifications of a state (for example, just what sort of liver involvement) and deciding between the various causes of a particular state. In context one task can often be solved in terms of the other (for example, of

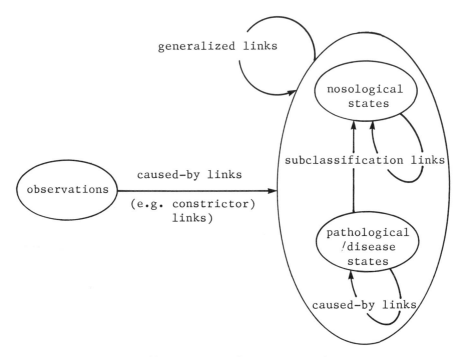

Fig. 6.1: Illustrating the conceptual structure of CADUCEUS' factual knowledge.

two causes one may involve an established nosological state and another may not, of two nosological states one may be caused by an established pathological state whilst another may not have any established causal antecedents). Accordingly there are links between the causal network and the nosological structure. These links serve to combine the two structures and reduce the apparent number of alternatives to be considered at any one decision point. Thus CADUCEUS obtains both a rapid focusing of attention and a critical assessment of possible attribution pathways of findings to their causes.

6.1 STATICS

The CADUCEUS' knowledge-base is more highly structured than the INTERNIST-I knowledge-base; it combines the advantages of a taxonomy of diseases and a causal network of pathological/disease states. Hence it provides a basis for combining two dimensions of diagnostic reasoning.

The disease taxonomy permits the sharpening of the system's focus of attention by exploiting easily observable cues (constrictors) in the patient data, whose presence is conclusive evidence for the presence of disease categories. The pathophysiological knowledge, in terms of underlying causal mechanisms explaining the attribution of observed manifestations to diseases, is necessary for the effective evaluation of proposed solutions to the diagnostic problems. These two basic knowledge structures are blended together via "generalized links". Figure 6.1 gives the basic conceptual structure of CADUCEUS' knowledge-base.

The causal network

Disease processes are modelled in terms of a causal network that links observations (symptoms, signs, laboratory data etc.) to their ultimate causes (disease entities) via intermediate pathological, or other disease states (e.g. syndromes), that form significant facets of the disease processes. The causal network is, therefore, structured to permit reasoning from effects to their ultimate (remote) causes.

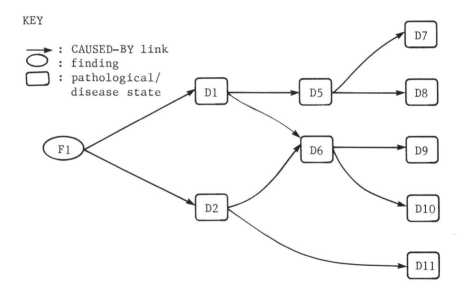

Fig. 6.2: Causal network depicting attribution pathways.

Referring to figure 6.2: An **attribution pathway** for some finding is a causal chain of pathological states linking the finding to some specific disease entity, and thus it constitutes an explanation for the finding in pathophysiologic terms. An attribution pathway for finding F1 is ⟨D1, D5, D7⟩. There are, in all, seven attribution pathways for F1 in the figure. The **differential diagnosis** list for some finding is the set of pathological/disease states that constitute the finding's immediate (proximate) causes. The differential diagnosis list for F1 is, therefore, the set {D1, D2}. Differential diagnoses are defined by intermediate pathological states as well, e.g. the differential diagnosis for D1 is {D5, D6}. Such **"proximate"** differential diagnoses should be compared with the **"remote"** differential diagnoses of INTERNIST-I. For example, the differential diagnosis list for F1, under INTERNIST-I, would have been {D7, D8, D9, D10, D11}.

The disease taxonomy

Groups of pathological/disease states (with no causal links between them) are clustered together to define nosological states. Such nosological states are further clustered together to define higher level nosological (e.g. clinical) states. The nesting of clusters (based on the concept of organ system involvement) gives rise to the disease taxonomy. The taxonomy is not strict, i.e. a given node can be clustered in more than one ways.

Generalized links

The two basic knowledge structures (causal network and disease taxonomy) are blended together via **"generalized links"** connecting pairs of nodes in the taxonomy: pathological states to other pathological states, pathological states to nosological states, or pairs of nosological states. The generalized links form a number of more abstract causal networks, representing the domain of internal medicine at the clinical level. Generalized links provide for a rapid convergence on tentative unifying hypotheses.

In figure 6.3 we show a number of generalized links which relate the causal network of figure 6.2 to a taxonomical structure. Note that disease state, D9, has been clustered under both nosological states N2 and N3.

Generalized links are of two types: one type of link asserts that every pathological state clustered under its antecedent is caused-by every pathological state clustered under its consequent (e.g. link D1 --➤ N5 in figure 6.3). The other type of link asserts that some pathological states clustered under its antecedent are caused-by some of the pathological states clustered under its consequent (e.g. link N4 --➤ N5 in figure 6.3).

Some of the generalized links are generated by propagating

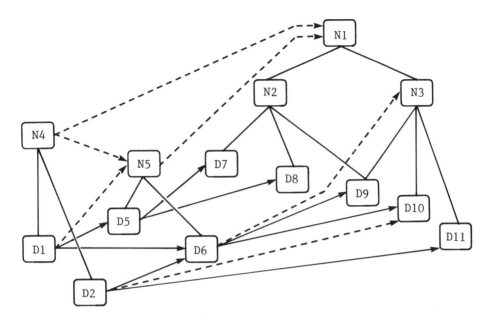

KEY:
⟶ : CAUSED-BY link; --▶: GENERALIZED link;
— : SUBCLASSIFICATION link; ☐ : pathological/
                                   nosological state.

Fig. 6.3: Taxonomies and generalized links.

Table 6.1	
Generalized link	Subpath chains
(N4 - -▶N1)	(N4 - -▶N5).(N5 - -▶N1)
	(D2 - -▶D10)
	(D2 ⟶D11)
(N4 - -▶N5)	(D1 - -▶N5)
	(D2 ⟶D6)
(N5 - -▶N1)	(D5 ⟶D7)
	(D5 ⟶D8)
	(D6 - -▶N3)
(D2 - -▶D10)	(D2 ⟶D6).(D6 ⟶D10)
(D1 - -▶N5)	(D1 ⟶D5)
	(D1 ⟶D6)
(D6 - -▶N3)	(D6 ⟶D9)
	(D6 ⟶D10)

low-level causal links upwards onto the disease taxonomies.
Chains of such links subsume low-level causal chains of the same
number of composite links. For example the high-level causal
chain N4 $\longrightarrow$ N5 $\longrightarrow$ N1 subsumes the seven possible attribution
pathways for F1. Other high-level links aggregate low-level
causal chains by spanning intermediate states. For example, the
high-level link N4 $\longrightarrow$ N1 is the abstraction of the above causal
chain. The interdependencies among generalized links are included
in the knowledge-base in terms of lists of subpath chains subsumed
by them. The interdependencies between the generalized links of
figure 6.3 are given in table 6.1.

Observations

Observations (historical items, symptoms, signs, etc.) are
linked to pathological, disease or nosological states. Some of
the findings are **constrictors**. The instantiation of a constrictor
linked to a nosological state results in the state being
concluded. For example, the observation of "jaundice" is
sufficient to warrant the conclusion of liver-involvement.

6.2 DYNAMICS

The goal of the system is to determine the set of specific
disease entities that collectively account for the patient's
illness. Pathological/nosological states constitute the
hypotheses, observations the direct pieces of evidence and
concluded states the circumstantial evidence.

The dynamics of the system can be abstracted in terms of the
hypothesis status transition diagram of figure 6.4. Initially all
hypotheses are **inactive**. A hypothesis is **activated** when it
becomes a member of the scope (decision-set) of some causal and/or
subclassification task/s (see text). When an element of the scope
of some task is **concluded**, i.e. when there is sufficient direct
evidence to warrant its conclusion, the remaining elements of the
scope return to the inactive status provided that they do not
belong to the scope of another, yet uncompleted, task. A
concluded hypothesis explains all those observations that can be
attributed to it. Every observed manifestation must be explained
by at least one concluded hypothesis.

Two dimensions of diagnostic reasoning

The structure of the knowledge-base allows for two orthogonal
types of diagnostic reasoning, each type corresponding to one of
the basic knowledge structures. The causal network permits the

construction of differential diagnoses from effects to causes
("left to right"), whilst the taxonomy enables the development of
differential diagnoses by successive refinements ("top down").
The left to right and top to bottom knowledge structures are
combined via the generalized links.

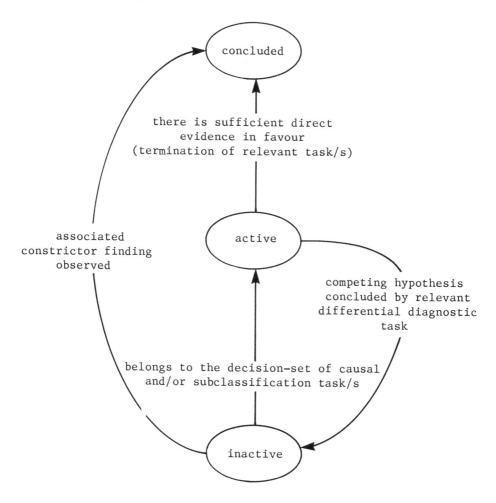

Fig. 6.4: Hypothesis status transition diagram.

Reasoning in a top down fashion is a characteristic human
reasoning process in medical problem solving (Elstein **et al,**
1978). Taxonomies permit the conceptualisation of a clinical
problem to be formulated in the most general terms, consistent
with the data, and the refinement of this conceptualisation takes
place as additional evidence is developed. High level nodes of

the taxonomy act to control the proliferation of active
hypotheses. In large taxonomies their value depends on the
availability of inherent mechanisms for focusing the diagnostic
process to the middle of them.

In CADUCEUS the focusing mechanism is provided via the
constrictor associations. A top to bottom differential diagnostic
task is set up at the point in the taxonomy determined by the
instantiation of a constrictor. Such a task is interpreted in
terms of a step-wise process where each step refines the
nosological state established in the previous step. The progression through the taxonomy narrows the set of feasible
solutions to a clinical problem by successively constraining the
set of outcomes compatible with the available clinical evidence.
The process stops at a terminal node. If the terminal node
corresponds to a pathological state, then its attribution pathway
needs to be established by the process explained below.

The causal network permits the construction of attribution
pathways from established effects (findings and pathological
states) to their remote or ultimate causes. An attribution
pathway is constructed by a series of differential diagnoses from
proximate to remote causes. For example, referring to figure 6.2,
the first step of the diagnostic process of deciding the
attribution pathway for F1, would be to decide between the
pathological states D1 and D2. Supposing that the decided state
were D1, then the next diagnostic step would be to decide between
the pathological states D5 and D6. Finally supposing that the
decided state were D5, then the final diagnostic step would be to
decide between disease states D7 and D8. If the decided disease
state were D8 then the decided attribution pathway for F1 would be
⟨D1, D5, D8⟩. In this procedure a situation can occur where a
given diagnostic step can not be decided without "looking-ahead"
in the causal network. The progression from one diagnostic step
to the next is usually associated with an increase in the
difficulty of acquiring evidence in deciding the new step. For
example, deciding between D1 and D2 is easier than deciding
between D5 and D6 which in turn is easier than deciding between D7
and D8.

The two dimensions of reasoning are combined by the provision
of links connecting them. The set of proximate causes of a
finding/pathological state fall under some node in the tax-onomy
and generalized links from the findings/pathological states to the
lowest such nodes are provided. Also causal links are abstracted
to connect pairs of nosological states, thus allowing the
construction of high level differential diagnoses. Hence, the
generalized links provide for a rapid convergence on tentative
unifying hypotheses while at the same time enabling access to as

much detail as is available in the underlying causal network.

6.2.1 Outlining the Overall Diagnostic Strategy

The overall diagnostic strategy is to find constrictors which take the problem solving activity deep into the taxonomy, thus constraining the possible solution set, and then to move between the differential diagnostic tasks of the two dimensions of reasoning to resolve this set. The elementary components of these differential diagnostic tasks are defined as follows:

A **subclassification task** for a concluded nosological state, decides which of the immediate forms (specialisations) of the concluded state is in fact present. The set of all immediate forms is the scope of the task. The termination of a subclassification task, marked with the conclusion of an element in its scope, might instigate a new subclassification task and/or a new causal task.

A **causal task** for an observed finding or a concluded pathological/nosological state, decides the "proximate" cause of the established effect. The scope of a causal task consists of the relevant set of possible proximate causes that can be reached via caused-by links, in the case of pathological states or observed findings, and via generalized links, in the case of nosological states. The termination of a causal task, marked with the conclusion of an element in its scope, might instigate a new causal task and/or a new subclassification task.

A number of elementary differential diagnostic tasks (of both types) are set up from those pathological/nosological states established on the basis of constrictor findings in the initial patient data (the constrictor concept is, therefore, exploited in the design for setting in motion a number of parallel or concurrent tasks). The concluded states and the set of states that fall under the scopes of the associated differential diagnostic tasks constitute the initial context for the diagnostic process. At every subsequent stage of the consultation there are a number of pending causal and subclassification tasks. The consultation terminates when every task is completed, with no decisions justified on heuristically imposed constraints. The set of concluded specific disease entities, collectively account for the patient's illness.

During the diagnosis there are multiple, not necessarily competing partial solutions to the clinical problem. These are maintained by the system with a single locus centered on a structure representing the initial context and other subordinate parts of the structure representing partial solutions (**composite** hypotheses). At every stage during a consultation the most promising partial solution is explored further until all the tasks defined on it have been completed. A partial solution can be explored in the following two ways:

The first way of exploring a partial solution space is by the heuristic synthesis of groups of differential diagnostic tasks defined on the space. Such heuristic synthesis operations function to constrain the scopes of the synthesized tasks into their most promising alternatives. Usually the range of alternatives is progressively constrained as more and more elementary diagnostic tasks are synthesized. Commonly, synthesis opportunities arise because independent diagnostic tasks share alternatives. The shared alternatives are taken as the most promising ones, forming the scope of the single, non-elementary, task resulting from the unification of the relevant tasks. Each synthesis operation applied on some space yields a new partial solution space subordinate to the given space.

The second way of exploring a partial solution space is by eliciting items of information that function to resolve the scopes of diagnostic tasks (elementary or not) defined on the space, thus bringing the immediate tasks to completion, and possibly creating new tasks. Any information acquired must be incorporated in every partial solution. If such information invalidates the synthesis manoeuvre that gave rise to a partial solution space then the given space is infirmed.

6.2.2 Focusing Mechanism

The efficiency of the design depends critically on the effectiveness of the focusing mechanism. The entire design is directed towards the improving of INTERNIST-I's simple focusing mechanism, revolving around the partitioning heuristic. Apart from the focusing effects of the constrictors, there are the following means for focusing the attention of the diagnostic process:

Generalized links and heuristic task synthesis

The **generalized** links, especially those that span large areas of the causal network provide, after the initial stage of the consultation, a primary mechanism for focusing the diagnostic process.

Referring to figure 6.5: The observation of finding F1 sets up the enementary causal task of resolving between its potential

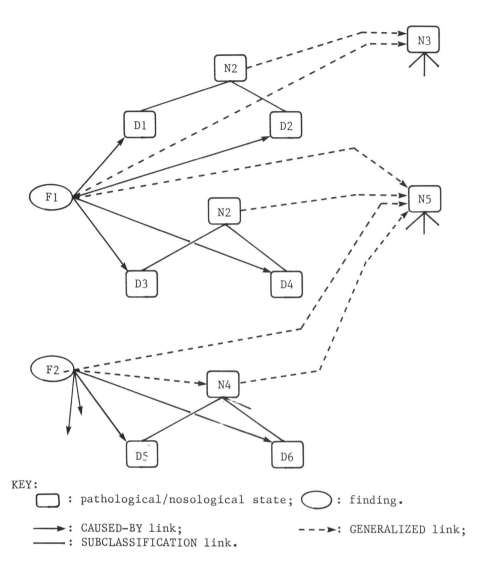

KEY:

☐ : pathological/nosological state; ⬭ : finding.

⟶ : CAUSED-BY link;   ---▶: GENERALIZED link;
⟶ : SUBCLASSIFICATION link.

Fig. 6.5: Generalized links as a primary focusing mechanism.

proximate causes D1, D2, D3 and D4. Generalized links connect F1
to nosological states N3 and N5 which collectively classify the
remote causes of F1. Supposing that the presence of N3 can be,
and is, easily established through constrictor observations, then
the tendency would be to focus the above causal task on D1 and D2
as the most promising explanations of the observation of F1. No
conclusion can be made about D1 and D2 without additional direct
evidence.

Generalized links, however, are mainly used for synthesizing independent diagnostic tasks. Suppose that findings F1 and F2 have been observed in the patient. Both F1 and F2 converge onto nosological state N5, making D3, D4 and D5, D6, respectively, the most promising explanations for the observations of F1 and F2.

Figure 6.6 depicts another possible heuristic synthesis manoeuvre, through generalized links, this time involving a causal task and a subclassification task. In this case the two elementary tasks can be replaced by a single non-elementary task of resolving between the alternatives B and E (the promise of alternatives, therefore, depends on the amount of circumstantial evidence, both causal and nosological, in favour of them); establishing one of these two alternatives would mark the termination of both elementary tasks. If, however, none of these two states is in fact present in the patient, then the above synthesis manoeuvre has been refuted and the elementary causal task should return to resolving between C and D. Similarly the subclassification task should return to resolving its remaining undenied alternatives.

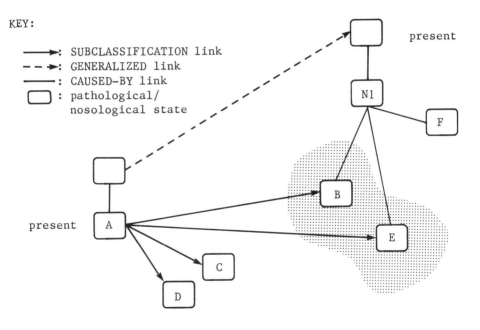

KEY:

⟶►: SUBCLASSIFICATION link
− − ►: GENERALIZED link
⟶: CAUSED-BY link
▢ : pathological/
      nosological state

Fig. 6.6: Synthesising causal and subclassification tasks.

Scoring of composite hypotheses (partial solution spaces)

At every stage of the consultation the diagnostic process focuses attention on the most promising composite hypothesis (partial solution space). The design specifies that any function employed for evaluating the promises of composite hypotheses must be based on Occam's razor, that states that the simpler of competing hypotheses is to be preferred to the more complex. Hence a function that makes the promise dependent on the number of elementary differential diagnostic tasks unified into a single complex task definition, would satisfy the above heuristic criterion.

Scoring of active pathological/nosological hypotheses

The CADUCEUS scoring function for evaluating the relative promises of the elements of the scope of the diagnostic task currently being explored is an improvement of INTERNIST-I's corresponding scoring function. The basic improvement is that the CADUCEUS scoring function is context dependent; each supplied finding is considered relevant for some nosological state if and only if there exists some pathological/disease state under the nosological state that has a strong association with the given finding. Active hypotheses classified under a nosological state are scored on the basis of those findings belonging to the context of the nosological state. This scoring function is outlined in Pople (1977).

6.2.3 Information Acquisition Strategies

The present design does not address this issue -- the INTERNIST-I information acquisition strategies for resolving decision-sets will eventually be interfaced to the design. It is hoped that the improved focusing mechanism would yield much more discriminating information gathering capability than was possible in INTERNIST-I.

Both the causal and subclassification tasks are basically information acquisition tasks, where the items of information elicited function to resolve between the elements of their scopes. Sometimes a causal/subclassification task is solved in terms of subclassification/causal tasks (see figure 6.7); this happens when the items of information required to solve the latter are more easily obtainable than those required to solve the former.

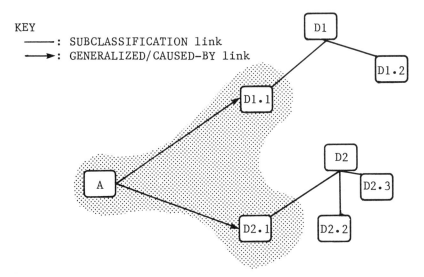

KEY
——: SUBCLASSIFICATION link
——▶: GENERALIZED/CAUSED-BY link

(a) Decide which of the causes of the established state, A,
is present by deciding which of the relevant forms are
present (nosological states can be ruled out if their
associated constrictors are absent).

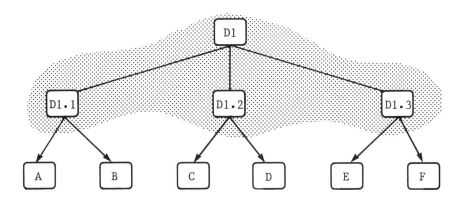

(b) Decide which of the forms of the established nosological
state, D1, is present by deciding which of the relevant
causes are present.

Fig. 6.7: Solving tasks of one type in terms
of tasks of the other type.

REFERENCES

Elstein A.S. Shulman L.A. and **Sprafka S.A.** (1978) : **Medical Problem Solving: an analysis of clinical reasoning,** Harvard University Press, Mass.

Pople H.E. (1977) : "The formation of composite hypotheses in diagnostic problem solving: an exercise in synthetic reasoning.", Proc. **IJCAI-77,** pp. 1030-1037.

Pople H.E. (1982) : "Heuristic methods for imposing structure on ill-structured problems: the structuring of medical diagnostics", in Szolovits P. (ed.) **Artificial Intelligence in Medicine,** AAAS Symposium Series, Boulder, Colo: Westview Press, pp. 119-185.

Chapter 7
CASNET

Application area: Medicine.

Principal researchers: S. Weiss and C. Kulikowski
(Rutgers University).

Function: Long-term management of diseases
whose mechanism is well known.

The CASNET system (Kulikowski, 1966; Weiss, 1974 and Weiss **et al**, 1978a) is in principle, a general tool for building expert systems for the diagnosis and treatment of diseases whose mechanisms are well known; diseases are modelled in terms of Causal **AS**sociational **NET**-works. In its major application, the diagnosis of, and the recommendation of treatment for, glaucoma, CASNET has exhibited experts' performance levels (Weiss, 1977 and Weiss **et al**, 1978b).

A survey of the current literature on CASNET showed that some of its features have not been implemented in any of its applications. However, even if the full CASNET model has not yet been demonstrated through some working system, nobody can claim full insight into the expert systems technology without having a good understanding of CASNET. By separating and making explicit the models of the diseases (factual knowledge) from the decision making strategies (reasoning knowledge), CASNET encaptures, or provides an efficient basis for the building of, important aspects which characterise expert consultant systems. These are: intelligent ordering of questions i.e. an aspect of natural dialogue structure, efficient explanatory facilities, efficient control over knowledge transitions, and an efficient basis for building teaching facilities.

OVERVIEW

CASNET's model of diseases is divided into three conceptually separated 'planes': 1) Observations 2) Pathophysiological states and 3) Disease categories. There is a fourth plane of therapy plans orthogonal to these three (see figure 7.1). The central plane is the pathophysiological one; disease processes are modelled by a network of pathophysiological states. Each link in the network has associated with it a numerical value representing a causality strength in proportion to the frequency that the first state causes the second. The disease categories are specified as tables of confirmed and denied pathophysiological states. Observations about a patient are used to confirm or deny particular states in the central network. These observations are made by the guidance of a network of history, signs, symptoms and laboratory tests nodes. These nodes are instantiated by information provided by the user: it may be that not all nodes will be instantiated. As observations are only more or less indicative of associated states, the associational links have a numerical value, representing a degree of indication, (or 'support'). Many observations will be more or less indicative of the same state, and an overall degree of direct evidential support for that state is calculated from the various degrees of support given by the observations. A state is marked as confirmed if it has a degree of support greater than a preset value. A state is marked as denied if its degree of support is less than a preset value. If neither confirmed nor denied a state is undetermined. Diagnosis is made by detecting pathways that traverse no denied states and identifying the associated disease categories.

7.1 STATICS

The knowledge-base contains information on observations, i.e. historical items, symptoms, physical signs and laboratory abnormalities, pathophysiological states, disease categories, therapy plans and their interrelationships.

7.1.1 Plane of Observations

There are relationships between observations that are used to establish local control over the sequence of questions in a fashion consistent with medical practice. These relationships are purely statistical or are based on phenomena constraints; they specify truth values of observations that can be directly deduced from already conducted observations.

Observations are direct evidence for hypothesised states. Hence the degrees of indication, Q, assigned to the associational links from observations to states can be readily interpreted as confidence measures of inverse inference.

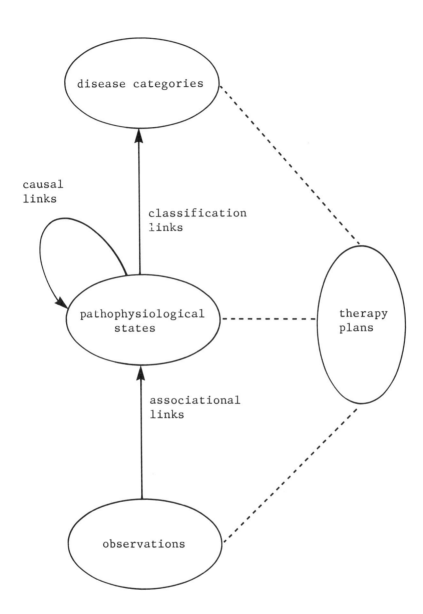

Overview of the CASNET model.

Fig. 7.1

The **ASSOCIATIONAL** relation from the domain of observations (tests) to the domain of pathophysiological states is defined by:

$$t_i \xrightarrow{\quad Q_{ij} \quad} n_j \;\; ; \;\; -1 <= Q_{ij} <= 1$$

where:

Q_{ij} is the confidence in state n_j given that result t_i is observed to be true. If Q_{ij} is a positive value then it denotes the degree to which observation t_i implies the presence of n_j (analogous to $P(n_j/t_i)$). If Q_{ij} is a negative value then $|Q_{ij}|$ denotes the degree to which observation t_i implies the absence of n_j (analogous to $P(\sim n_j/t_i)$). Associated with each observation are costs C_i^j that reflect the cost of obtaining the result t_i.

7.1.2   Plane of Pathophysiological States

The mechanisms of a disease are described in terms of a causal network of pathophysiological states. States are not diseases but detailed dysfunctions. Some of the pathophysiological states have no dysfunctional consequences (**final** states) and these are linked by causal pathways to primary pathophysiological states (**starting** states). A complete disease process is a complete pathway through the causal network from a starting to a final state. A partial pathway from a starting state to a non-final state represents the degree of evolution within the disease process. Progression along a causal pathway is usually associated with increasing seriousness of the disease. However, the actual resolution of the desease process into a series of states linked by a strength of causation is a design consideration to help guide the process of diagnosis; it is not the case that the pathophysiological plane in CASNET is a deep causal model of the disease development. Many events and complex relationships may be summarised by characterizing them as a 'state'.

A simple example (c.f. Weiss **et al**, 1978a) of a network of states is given in figure 7.2, where n_i is a state and a_{ij} is the degree to which state n_i causes n_j. This strength of causation corresponds to qualitative ranges such as: sometimes, usually, often, always. Starting states have prior or starting frequencies based on their relative frequencies of occurrence.

It is not required that a set of mutually exclusive and exhaustive effects be defined for every cause, nor a similar set of causes be defined for every effect, but the net is not allowed to contain loops. The assigned strengths of causation are independent of the manner in which a state is itself produced and

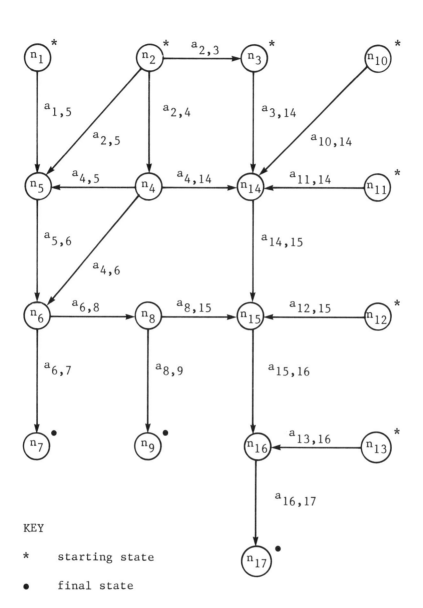

KEY

*   starting state

●   final state

State Network.

Fig. 7.2

there are no constraints on the entire set of strengths of
causation emanating from a single cause.

The **CAUSAL** relation on the domain of pathophysiological
states is defined by:

$$n_i \xrightarrow{a_{ij}} n_j \quad ; \quad 0 <= a_{ij} <= 1$$

Through the causal relation, a network of states is defined.
The formal definition of a network of states is a quadruple of (S,
F, X, N) where S is the set of starting states, F is the set of
final states, N is the total number of states and X is the
extension of the causality relation.

7.1.3 Plane of Disease Categories

Several causal mechanisms (starting with the starting states
and finishing on a final state) may be included in the same
disease category. Thus a disease category is definable through
its final state by the subsumption of the ordered sequences of
states that lead from starting states to it.   A subcategory is
defined in terms of some important intermediate (treatable) state
of the sequences. Thus subcategories indicate severity of the
progression of a disease.

Referring to figure 7.3, disease category D_1 is the three
ordered sequences of states $\langle n_1,n_3,n_5,n_6 \rangle$, $\langle n_2,n_3,n_5,n_6 \rangle$ and
$\langle n_4,n_5,n_6 \rangle$. Similarly, category D_2 is the two ordered sequences
$\langle n_1,n_8,n_9,n_{10} \rangle$ and $\langle n_7,n_9,n_{10} \rangle$. Suppose that n_3 and n_5 are
important intermediate states of the sequences in disease category
D_1, then the subcategory D_{12}, corresponding to intermediate state
n_5, is the three sequences $\langle n_1,n_3,n_5 \rangle$, $\langle n_2,n_3,n_5 \rangle$ and $\langle n_4,n_5 \rangle$,
the subcategory D_{13}, corresponding to intermediate state n_3, is
the two sequences $\langle n_1,n_3 \rangle$, $\langle n_2,n_3 \rangle$. Similar subcategories can be
defined for D_2 where, say, the important intermediate states are
n_8 and n_9.

The **CLASSIFICATION** relation is a boolean relation from the
set of complete disease processes (ordered sequences of states,
from starting to final states) to the set of disease categories.
This is defined by:

$$\langle n_s, \ \ldots \ n_j, \ \ldots \ n_f \rangle \longrightarrow D_i$$

(the ordered sequence $\langle n_s, \ \ldots \ n_j, \ \ldots \ n_f \rangle$ implies disease
category D_i)

The data structure used for holding the information on a

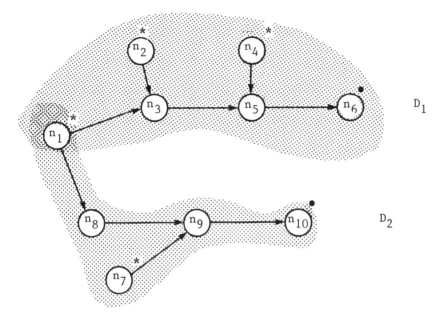

Fig. 7.3: Disease pathways.

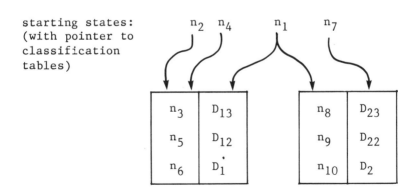

Fig. 7.4: Classification tables.

The disease category "Primary open angle glaucoma",
D_1, say, is implied by the following pathway:

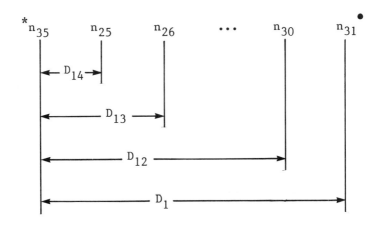

where: D_{14} = mild risk of open angle glaucoma
       D_{13} = high risk of open angle glaucoma
       D_{12} = very high risk of open angle glaucoma;
                  significant risk of visual field loss

       D_1 = primary open angle glaucoma

The relevant classification table
augmented to include therapy plans is:

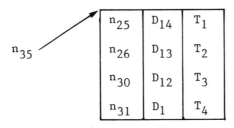

where: T_1 = return visit in 6 months
       T_2 = careful follow up with repeated tension readings
       T_3 = careful follow up or a therapeutic trial with
               pilocarpine 1% Q.I.D.
       T_4 = miotic therapy

Fig. 7.5(a)

disease category is known as a classification table (one table for each category).

Each disease category has pointing to its classification table all the starting states that begin the sequences that define it. Figure 7.4 gives the classification tables of the disease categories conceptually represented in figure 7.3.

7.1.4 Plane of Therapy Plans

The knowledge dealing with recommendation of treatment is structured on a distinct plane that contains the mechanisms of therapy. This plane can be visualized as being orthogonal to the other planes because it encompasses a qualitatively different domain, that of actions. The plane of therapy plans is primarily related to the plane of the disease categories, but it is also associationally related to the plane of observations. Such associations represent indications and contra-indications, of various strength, to particular treatments within the therapy plans.

The relations beween disease categories, subcategories and therapy plans are stored by extending the classification tables for the disease categories. We demonstrate this by a concrete example in figure 7.5(a) (c.f. Weiss et al (1978a)).

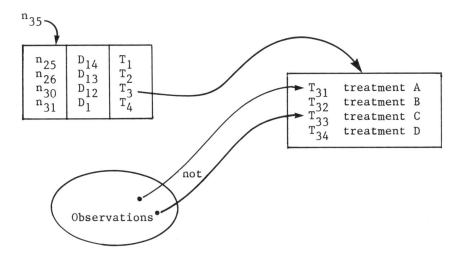

Fig. 7.5(b): Treatment recommendations are related to both disease categories and observations.

It may be the case that a therapy plan, T_i, is an ordered category of treatments T_{i1}, T_{i2}, ..., T_{im}. The order is taken as a default preference of treatment for that disease category. In certain situations no well established set of preferences of treatment exists and selection is based on the pattern of given observations. These patterns are defined by another associational relation from the domain of observations to **particular** treatments (see figure 7.5(b)).

The **ASSOCIATIONAL** relation between observations and particular treatments is defined as follows:

$$t_i \xrightarrow{\text{Pf}_{i(kj)}} T_{kj} \quad ; \quad -1 <= Pf_{i(kj)} <= 1$$

where:

$Pf_{i(kj)}$ is the preference of treatment T_{kj} given that t_i is observed. If $Pf_{i(kj)}$ is a positive value then it denotes the degree to which observation t_i favours the application of treatment T_{kj}. If $Pf_{i(kj)}$ is a negative value then it denotes the degree to which observation t_i disfavours the application of treatment T_{kj}.

7.2 DYNAMICS

The diagnostic process in CASNET is, basically, to infer the disease processes that are taking place within the patient, i.e. to determine which part of the causal network is actually operative in the patient. Observations constitute pieces of direct evidence, pathophysiological states figure as intermediary hypotheses and causal pathways figure as higher level hypotheses.

7.2.1 Outlining the Overall Diagnostic Strategy

The overall diagnostic strategy will now be outlined. At the beginning of a consultation the user either enters a set of initial observations or responds to a preliminary sequence of questions. The purpose of this initial questioning is to home in quickly on the problem area and thus to reflect the practice adopted by human experts. These initial observations instantiate some of the associational links, and the confidence factors (defined below) of the relevant states are updated (hypothesising pathophysiological states) and thus the initial configuration of status values is determined. The possible potential causal pathways explaining these observations are determined (hypothesising operative causal pathways) and the undetermined states on

these pathways, most likely to be confirmed or infirmed, are
selected for further exploration. Subsequent questioning is
directed to these states. The elicited observations instantiate
other associational links and the cycle of updating confidence
factors, hypothesising operative causal pathways, and eliciting
further direct evidence, is repeated until no more questioning is
judged to be fruitful. At the end of the cycle the most likely
explanation of the disease processes manifested in the patient is
determined by identifying the set of most likely starting states.
Control is then passed to the therapy selection strategy.

**Hypothesising pathophysiological states
from given observations**

Each state in the pathophysiological plane may have several
observations associated with it. The link with the strongest
associational value determines the overall confidence that the
state is present or not present in the patient. This
associational strength becomes the confidence factor (CF) assigned
to the state. Should the CF be greater than or less than a preset
value then the state is marked as confirmed or denied. The
undetermined states (states neither confirmed nor denied) have
their CFs updated as new observations are made.

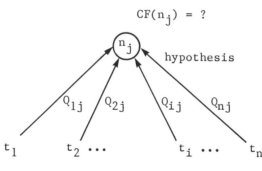

Fig. 7.6: Assigning confidence factors to states.

The question posed by figure 7.6, is answered formally as
follows:

Initially the CF of all states is $CF(n_j) = 0$. When some
observation t_i is made and t_i is associationally related to state

n_j, the Q_{ij} value of t_i determines the CF of n_j by the update
procedure:

    i)   If $|CF(n_j)| < |Q_{ij}|$ then $CF(n_j) := Q_{ij}$;

    ii) If $CF(n_j) = -Q_{ij}$ then $CF(n_j) := 0$; conflict noted;

    iii) Otherwise $CF(n_j)$ is unchanged.

In case ii), where the confidence in the presence of n_j equals the
confidence in its absence, a conflict is said to exist.

Thus according to the updating steps i)–iii) the overall CF
of n_j is given by:

    a) $CF(n_j) = \max_{t_i} \{|Q_{ij}|\} = |Q_{kj}|$

    b) If $Q_{kj} < 0$ then $CF(n_j) := -1 * CF(n_j)$.

In other words only the result with the strongest associational
strength is accounted for.

From the certainty factors the STATUS of states can be
determined as follows:

Let   be a positive threshold value.

    i) If $CF(n_j) >= \Theta$ then STATUS $(n_j) :=$ 'confirmed';
    ii) If $CF(n_j) <= -\Theta$ then STATUS $(n_j) :=$ 'denied';
    iii) Otherwise STATUS $(n_j) :=$ 'undetermined'.

The basic status transition graph is given as figure 7.7. In
this figure the undetermined status has been refined into four
classes: dormant, partially confirmed, partially denied and
inconsistent. Initially all states are dormant. At the end of
the consultation some states will be confirmed, some denied and
some will remain undetermined.

It is interesting to compare this model for assigning
confidence factors to the probabilistic model employed by
PROSPECTOR. To start with there is a significant difference
between CASNET's associational strengths, Q_{ij}, and conditional
probabilities. For example from $Q_{ij} = 0.3$ (that is, there is a
confidence of 0.3 that n_j is present given observation t_i) we can
not deduce that there is a confidence of 0.7 that n_j is absent;
observation t_i counts either towards the presence or absence of
state n_j, not both (CASNET is similar to MYCIN in this respect).

KEY: - - - - conflict resolution
         ———— update

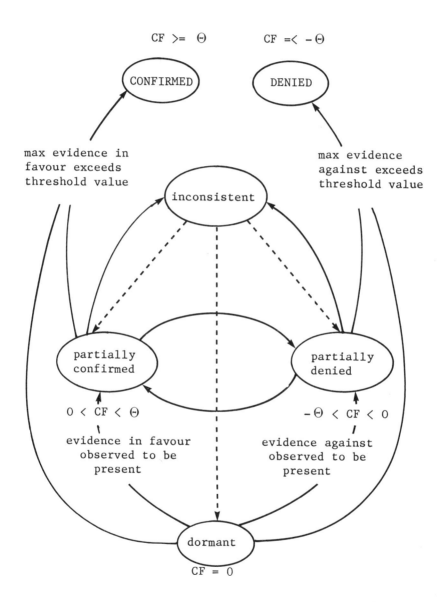

Fig. 7.7: State status transition diagram.

Therefore associational strengths do not obey the law of
probabilities that states $P(n_j/t_i) + P(\sim n_j/t_i) = 1$. The
probabilistic model assumes independence between the individual
pieces of evidence even when the violation of this assumption is
obvious. No such assumption is required for the CASNET. Finally,
the probabilistic model allows uncertainties to be associated with
pieces of evidence. In CASNET a result is true, false or
undetermined (a three-valued logic can be used to determine the
value of a complex result).

Hypothesising operative causal pathways

A particular configuration of status values over the causal
network generates constraints on the hypothesis of causal pathways
(disease processes) which are operative in the patient. A
hypothesis is precisely a class of causal pathways that explain
the occurrence of the observations. To define the relevant types
of pathways we first give the definition of an 'admissible
pathway' as "a pathway that contains no denied states".

The set of potential explanations of the observations can be
obtained by generating, from the undenied starting states, all the
admissible pathways that have at least one confirmed state and
terminate on a confirmed state (these may be the same state). The
most likely hypothesis, however, is defined by the set of most
likely starting states. This set of starting states is the
minimal subset of undenied starting states from which it is
possible to generate admissible pathways which end on a confirmed
state and, collectively, contain all the confirmed states.

In these definitions of most likely and potential explanations the pathways are terminated on confirmed states. Con-
tinuing these pathways beyond these points to include all
undetermined states, **possible current** aspects of the disease
process are indicated. Possible **future** developments of the
disease process may be obtained by further extending possible
pathways to final states even though they then may traverse
currently denied states; such states may be **predicted** to be
confirmed on a **prognostic assessement** of the likely disease
progression.

Referring to figure 7.8: The set of **most likely starting
states** consists of starting state n_1 only, because the pathways
$\langle n_1,n_2,n_3,n_4,n_5,n_8,n_{13}\rangle$ and $\langle n_1,n_{13}\rangle$ generated from it directly
explain the current observations. Thus they collectively con-
stitute the most likely hypothesis. The **potential explanations** of
observations are the causal pathways $\langle n_1,n_2,n_3,n_4,n_5,n_8,n_{13}\rangle$,
$\langle n_1,n_{13}\rangle$, $\langle n_1,n_2\rangle$ and $\langle n_{10},n_{13}\rangle$. Extending the pathways which
constitute the most likely hypothesis we indicate that a **possible**

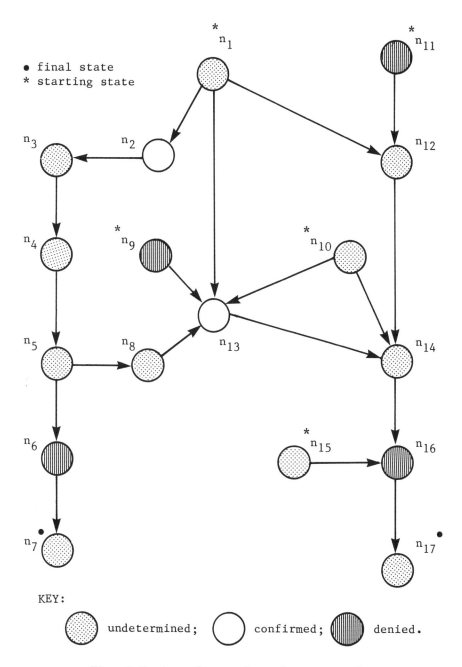

Fig. 7.8: A configuration of status values.

aspect of the disease process is the occurrence of state n_{14}.
Similarly we can indicate a possible aspect of the **potential**
disease process $\langle n_1, n_2 \rangle$ is the pathway $\langle n_1, n_2, n_3, n_4, n_5 \rangle$.
Extending the possible pathway $\langle n_1, n_2, n_3, n_4, n_5, n_8, n_{13}, n_{14} \rangle$ to the
final state n_{17}, we traverse the denied state n_{16}; this currently
denied state (n_{16}) may be **predicted** to be confirmed on a
prognostic assessement of the likely disease progression.

Conflicts and contradictions

The diagnostic process must be in a position to detect and
handle the following conflicting/contradictory situations that can
arise during a consultation:

1)   A conflicting situation occurs when the confidence in the
      presence of a state equals the confidence in its absence.   In
      such situations the state is classified as undetermined,
      until additional observations resolve the conflict.

2)   A contradictory situation occurs when a state is confirmed,
      yet all of its potential causes in the network are denied.
      This situation might have the explanation that the model of
      the disease is incomplete, i.e. some causes of the state have
      been left out.   If this is the case, the detection of the
      contradiction is a means of detecting missing knowledge, thus
      leading to knowledge-base updates.   Otherwise the situation
      must be resolved reevaluating the observations.

3)   If some undetermined state on a pathway that forms part of
      the most likely explanation of the observations is "inclined"
      towards its denial then the pathway is inconsistent.   This is
      resolved either by eliciting additional more conclusive
      evidence or by seeking an alternative explanation of the
      observations, which while not the most likely, is entirely
      consistent with the states that are explained.

7.2.2   Information Acquisition and Focusing Strategies

A low-level, but effective, information acquisition strategy
requires the questions for the consultation to be organized to
reflect the inherent structure of the areas investigated.   Within
each area, questions are grouped together and entered hierarchically.   Thus patterns of questions may be sequentially
entered by using these local area tree structures.   Also, it may
be the case that any particular question has a condition or set of
conditions that have to be met before it can be asked
intelligently.   If the specified conditions are not met, the
question is not asked.   For example, "the cup-to-disc ratio at the

optic nerve head can be determined only if an opthalmoscopic examination is performed".

In addition to the low level strategy CASNET has two higher level strategies that have intimate connection to the focusing aspects of the diagnostic process. Before we consider these strategies we must emphasise that in CASNET's diagnostic process the notion of competing hypotheses does not exist, either at the level of intermediate state hypotheses or at the level of causal pathways hypotheses. CASNET permits the coexistence, in the same patient, of multiple causes/effects of some state and thus the coexistence of a number of disease processes, i.e. the same observation can have a number of explanations. Thus CASNET only focuses its diagnostic process with the objective of confirming or denying a state -- independently of all the other hypothesised states -- and not with the objective of confirming one state at the infirmation of another. Thus it is unlike systems like INTERNIST-I, CADUCEUS and ABEL whose focusing and information acquisition strategies are concerned with identifying and resolving between competing hypotheses.

The diagnostic process must focus its attention on the most promising part of its search space. In CASNET this means determining the most promising part of the causal network, i.e. determining those causal pathways that constitute potential explanations of the current observations. Undetermined states on these pathways suggest investigations. Referring to figure 7.8, two possible potential causal pathways are $\langle n_1, n_2, n_3, n_4, n_5 \rangle$ and $\langle n_1, n_{13}, n_{14} \rangle$. This suggests that the states to be explored next should be states n_3 and n_{14}. Exploring a state means finding out from the user direct evidential information; such information it is hoped will lead to the confirmation/denial of the state.

Circumstantial causal evidence.

An alternative focusing strategy to that of selecting the most promising pathway and then a state on that pathway is to select states directly. However, as the method we detail below can also be used as a method of selecting a promising state on a promising pathway, it may be used in combination with a path selection strategy.

Attention may be focused on a state by considering cicumstantial causal evidence. The general nature of this reasoning may be outlined simply as follows:

An event or state of affairs is likly to occur if one of its possible causal antecedents has occurred. Moreover, the greater

the number of its possible causal antecedents that have occurred
the more likely it is to occur. On the other hand, the occurrence
of an event or state of affairs lends weight to the likelihood
that one of its causal antecedents has occurred. Although the
likelihood of a particular antecedent having occurred is not
greater on the evidence that it is also a possible causal
antecedent of another event or state of affairs which has
occurred. Futhermore, that an event or state of affairs has not
occurred is evidence that none of its causal antecedents has
occurred. An event or state of affairs that is both the
antecedent of one event or state of affairs and the consequence of
another is not rendered extra likely because both of these occur;
it is as likely as the strongest evidence.

In CASNET the confirmation and denial of states in the
network provides the basis of causal evidence to focus attention
to a state which is likely to be confirmed or denied. This
evidence is captured in a likelihood measure. Each state has
associated with it a numerical value (a **weight** not the CF) which
is interpreted as a likelihood measure of the degree to which the
state is expected to be confirmed or denied from the evidence
provided by the pattern of confirmed and denied states in the
network. Evidence from **confirmed** causal antecedents is cumulated
(forward causal evidence) and evidence from **confirmed** causal
consequents is selected (backward causal evidence). The strongest
of these is taken to give the weight of the state, i.e. the weight
of a state is the maximum of the forward and backward evidence.

The mechanism for assigning WEIGHTS to states

The strategy of choosing to investigate a state with a high
(or low) weight is equivalent to using causal evidence reasoning
(because the weight encapsulates this reasoning) to focus the
overall diagnostic strategy.

The WEIGHT, W_i, assigned to state n_i is given by:

$$W_i = \max \{ w_F(n_i), w_B(n_i) \}$$

where:
$\quad\quad w_F(n_i)$ is the forward weight assigned to n_i,
$\quad\quad w_B(n_i)$ is the backward weight assigned to n_i.

The **forward weight**, $w_F(n_i)$ represents the likelihood that
state n_i will be eventually confirmed, from evidence provided by
its confirmed causes.

$\quad\quad w_F(n_i)$ is given by:

$$w_F(n_i) = \sum_j w_F(i/j) * \mu_j$$

where:

we take the admissible pathways generated from the nearest confirmed or starting states and let j range over the initial states of this set (n_j may be the initial state of more than one pathways).

$$\mu_j = \begin{cases} \text{the prior or starting frequency, } a_j \text{ of } n_j \\ \text{when } n_j \text{ is an unconfirmed starting state.} \\ \\ 1 \quad \text{when } n_j \text{ is a confirmed state.} \end{cases}$$

$w_F(i/j)$ is the likelihood that n_i will be brought about given that its causal antecedent n_j is confirmed, given by:

$$w_F(i/j) = \prod_{k=1}^{i-1} a_{k,k+1}$$

i.e., $w_F(i/j)$ is the product of the link strengths between all pairs of successive states (n_k, n_{k+1}) in the pathway from n_j to n_i. This means that in the simple case the likelihood of a state on the evidence of a confirmed antecedent is equal to the strength of causation.

Hence all $w_F(i/j)$ values for all (n_i, n_j) pairs are obtained by computing the transitive closure of the causality relation X. Thus when n_i is a starting state·$w_F(n_i)$ equals n_i's prior frequency a_i.

The designers justify the use of summation in the mechanism for estimating the forward weight by saying that the states most likely to be present in the patient are those which have many possible ways of occurring.

The confirmation of a state increases the forward weight of all of its effects. Similarly, the forward weights resulting from a denied state n_i, will decrease since n_i cannot lie on any admissible pathway to another state.

The **backward weight**, $w_B(n_i)$, for state n_i represents the likelihood that n_i will be eventually confirmed, from evidence provided by its confirmed effects. $w_B(n_i)$ is given by:

$$w_B(n_i) = \max_j \{w_B(i/j)\}$$

where:

j ranges over the set of confirmed states that can be reached from state n_i.

> $w_B(i/j)$ is the likelihood that n_i will be confirmed given
> that its causal consequent n_j is confirmed. The formula for
> deriving it is analogous to Bayes' formula for inverse
> probability:
>
> $$w_B(i/j) = \frac{w_F(j/i) * w_f(i)}{w_f(j)}$$
>
> where $w_f(i)$ is calculated as $w_F(i)$ except that it is
> accumulated from starting states, i.e. confirmed states
> are ignored.

> Because an admissible pathway cannot contain a denied
> state, $w_B(i/j)$ is proportional to the weight of pathways
> passing through n_i to n_j, divided by the weight of all
> currently possible pathways to n_j.

The designers justify the use of maximum for obtaining the
overall backward weight by saying that the network should be
searched for strong evidence that n_i is present. Since several
confirmed states may in fact be unrelated, an average or sum would
appear to be less appropriate.

> The backward weight of a state may be increased when its
> effects are confirmed.

> As mentioned above the strategy of selecting the state with
> the highest weight is equivalent to selecting the state from
> circumstantial causal evidence. This strategy can be modified to
> take into consideration the costs of making an observation. (The
> notion of "cost" here could be expanded to include such notions as
> the fact that making an observation is dangerous to the patient.)
> For example, a less likely state associated with "cheap"
> observations may be preferable to a more likely state associated
> with "expensive" observations. Weiss **et al** (1978a), specifies two
> such heuristics.

Let: $t_i \longrightarrow n_j$;
    W_j denote the weight assigned to state n_j;
    C_i denote the cost of obtaining result t_i.

The two heuristics may be stated as follows:

> i) Max weight-to-cost ratio:
>   select t_i such that $W_j/C_i = \max_{m,n} (W_m/C_n)$

> ii) Max weight within a certain range of costs:

select t_i such that $W_j = \underset{m}{\max}(W_m)$ for all t_n with $C_n < C$

Questioning stops when no state has a weight that exceeds a preset threshold value. Setting the threshold value to zero gives an exhaustive search of the network.

7.2.3 Therapy Selection Strategy

As observations are received associational links between test results and particular treatments are instantiated, and the preference measures, PF, for these treatments are updated using the same rules as those for updating the confidence factors for pathophysiological states.

Finally the set of most likely starting states is used to access the classification tables. The entry selected from every accessed classification table is the deepest one containing a pathophysiological state that is confirmed for the patient. A specific treatment, T_{kj} say, is then selected from the corresponding therapy plan, T_k, according to the rules:

i) Select T_{kj} such that $PF(T_{kj}) = \underset{i}{\max} \{PF(T_{ki})\}$

ii) If there is more than a single treatment with maximum PF, select the one with smallest index j in the a priori prototypical ordering.

7.3 FACILITIES

7.3.1 Explanatory Facilities

CASNET is able to explain its diagnostic reasoning. Kulikowski (1966) briefly mentions its explanatory facilities. These are:

i) It can explain why a given state was confirmed or denied (evidence for states).

ii) It can explain why a given causal pathway was selected (likely causal pathways).

iii) It can explain how conflicts of evidence were resolved.

7.3.2 Knowledge Acquisition Facilities

The knowledge-base of the CASNET/glaucoma system is automatically updated by a group of ophthalmological clinical researchers who can access the system through the Ophthalmological NETwork (ONET). Such updates are performed through a separate editing program (Weiss and Kulikowski, 1973), that checks the updated knowledge-base for consistency. The editing program can be interfaced to any CASNET-based system. Also, because new results are included in the knowledge-base as soon as they are produced by clinical researchers, the system's capability as a teaching tool as well as a consultant is enchanced. Currently (1983), the CASNET/glaucoma knowledge-base contains more than 100 pathophysiological states, 400 test results, 75 classification tables and 200 diagnostic and treatment statements.

7.4 IMPLEMENTATION DETAILS

CASNET/glaucoma is running in 35K words of memory on a DEC 10 or 20 computer under either the TOPS-20 or TENEX operating systems. Because of speed and efficiency considerations, it is written in FORTRAN. The editing program is written in SNOBOL.

7.5 INCREASING THE MODEL'S RANGE

The CASNET model is effective for those problem domains which can easily be modelled in causal terms. According to the designers of the model many domains thou n do not easily fit into a causal framework; even for those domains which seem amenable to causal modelling, the effort necessary to describe the causal model can be quite extensive. The CASNET model has been applied, with considerable success, to the domain of glaucomas. It has also been applied to the domains of aneamias, thyroid dysfunction, diabetes and hypertension. It is thus relevant, only to those narrow medical domains whose pathophysiology is well understood. This led the designers of CASNET to produce a more general model, EXPERT (Weiss and Kulikowski, 1979). Although designed indepedently of any specific application EXPERT has been influenced by the designers' experience in developing consultation models in medicine and chemistry and simple prototype models such as those for automobile repair. The EXPERT system has been applied to medical domains of endocrinology, ophthalmology, and rheumatology and currently its application to domains outside medicine is being investigated. The CASNET and EXPERT models are compared and contrasted in Weiss and Kulikowski (1982).

REFERENCES

Kulikowski C.A. (1966): "Computer based consultation systems as a teaching tool in higher education", Report CBM–TR–66, Department of Computer Science, Rutgers University.

Kulikowski C.A. and Weiss S. (1973): "An interactive facility for the inferential modelling of disease", Proc. Princeton Conf. on Information Sciences and Systems, p. 524.

Weiss S.M. (1974): "A system for model–based computer–aided diagnosis and therapy", Ph.D Thesis, Computers in Biomedicine, Department of Computer Science, Rutgers University, CBM–TR–27–Thesis.

Weiss S.M. (1977): "A model–based consultation system for the long-term management of glaucoma", Proc. IJCAI–77, pp. 826–832.

Weiss S.M. and Kulikowski C.A. (1979): "EXPERT: A system for developing consultation models", Proc. IJCAI–79, pp. 942–947.

Weiss S.M. and Kulikowski C.A. (1982): "Representation of expert knowledge for consultation: The CASNET and EXPERT projects", in Szolovits P. (ed.), Artificial Intelligence in Medicine, AAAS Symp. Series, Boulder, Colo, West View Press, pp. 21–55.

Weiss S.M., Kulikowski C.A., Amarel S. and Safir A. (1978a): "A model–based method for computer–aided medical decision–making", Artificial Intelligence, Vol. 11, pp. 145–172.

Weiss S.M., Kulikowski C.A. and Safir A. (1978b): "Glaucoma consultation by computer", Comput. Biol. Med., Vol. 8, pp. 25–40.

Chapter 8

ABEL

Application area :	Medicine.
Principal researchers :	W.B. Schwartz (Tufts University / New England Medical Centre), P. Szolovits and P.S. Patil (Massachusetts Institute of Technology).
Function :	Diagnosis of electrolyte and acid-base disturbances.

The **A**cid-**B**ase and **EL**ectrolyte system (ABEL) (Patil, 1981) and Patil **et al** 1982a) partially implements the diagnosis module and the patient-specific model component of a paper system for the task of patient management. The paper system figures in an active research programme, and has, in addition to ABEL's components, a Decision module and a Therapy module.

OVERVIEW

ABEL's design was founded on insights gathered from re-evaluation of the techniques used in previous programs. Accordingly, one can overview the system by describing its promising features. These are that it deals with the knowledge of a disease phenomenon at different levels of detail. That it exploits the notion of causality in several ways: to organize the patient facts and disease hypotheses, to deal with the effects of more than one disease present in a patient and to provide the basis of explanations. That it captures the notions of adequacy and simplicity of a diagnostic posibility and hence has the possibility of not requiring numeric belief measures as criteria for diagnostic reasoning.

8.1 STATICS

The representation of the relevant medical knowledge in the ABEL system is based on the designers' belief that experts derive their competence from their ability to reason both with commonly occurring associations of syndromes and diseases (phenomenological knowledge) and with the best available pathophysiological (causal) knowledge about disease mechanisms. Phenomenological knowledge is necessary for efficient diagnosis exploration. Pathophysiological knowledge is necessary for proper understanding of a difficult case; it is frequently used in the evaluation and justification of diagnostic hypotheses.

8.1.2 Multi–level Description of Causal Knowledge

ABEL's knowledge-base consists of a causal network of nodes representing the domain of acid-base and electrolyte disturbances at the pathophysiological level. Nodes represent various types of states and quantifiable parameters. Nodes with no causal antecedents are the **ultimate etiologies**. In the domain of acid-base and electrolyte disturbances feedback loops are common and so the causal network contains loops.

Some of the **primitive** nodes of the pathophysiological level act as **foci** (landmarks) at a higher level of abstraction. This level of abstraction, containing the focus nodes of the lower level, may be considered to be a network of states where the nodes and links subsume the entire network of lower order nodes and links. On this level certain nodes may have focus node status and thus define another level of abstraction and so on. These levels above the pathophysiological level are not explicitly represented but pathways through them can be dynamically instantiated by a mechanism of using operations on **focus links** and focus nodes. Focus links, can be thought of as status designators: where a node has a focus "link" attached, it is a focus node and represents a region of the network at a higher level. At the higher level it can be thought of as a composite node subsuming the lower region. The composite node may, and often does, have the same name as the focus node at the lower level. At the pathophysiological level a node may be designated a focus status at each level of abstraction, i.e. it may have a chain of focus links attached, thus defining and linking each level (see figure 8.1).

Higher level links

Not all links in the network are interpreted as causal relations, some are merely associational relations or grouping

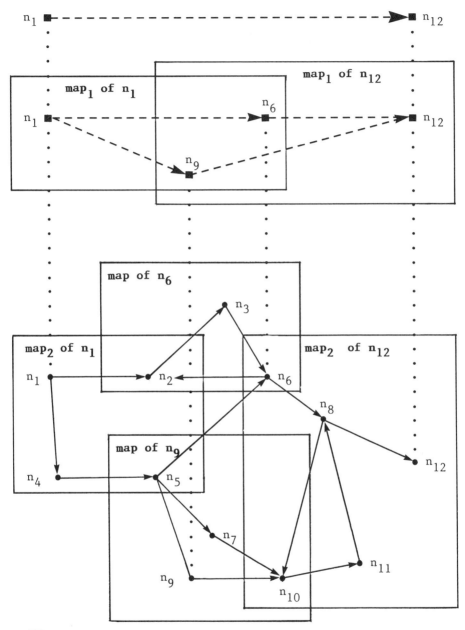

KEY:

⟶ :primitive causal link; – – – :subsumptive causal link;
• • • :focal link; ● : primitive node;   ■ :composite node.

Fig: 8.1: Multi-level description of causal knowledge.

links clustering together findings whose co-occurrence should be routinely recognized, e.g. findings that are strongly suggestive of syndromes. Thus the network is more like an associative network whose links are prominently causal.

The relations that are interpreted causally, usually represent strong associations not logical truths. The causal strength of a pathway is the product of the link strengths between the successive nodes, hence the longer a pathway, the more its causal strength degrades. The soundness of an inference degrades with every additional intermediary link.

The multi-level description of ABEL allows the aggregation of the diagnostic space into higher levels of abstraction. At any higher level each link covers large distances by subsuming commonly occuring pathways at the lower level and thus minimizes the possible errors of inference. Referring to figure 8.2:   The causal link ■ ----▶■ subsumes the pathways $\langle n_1, n_2, n_3, n_6 \rangle$ and $\langle n_1, n_4, n_5, n_6 \rangle$.

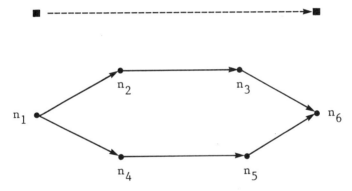

Fig. 8.2: Enhancing the soundness of inference.

There are situations where the fixed number of levels of descriptions limits the ability to aggregate causal descriptions. In this case the intermediate nodes in a causal pathway can not be subsumed, and a different link -- the **compiled** link -- represents the pathway. The compiled link allows the deeper exploration of a link without the degradation of the soundness of inference. Referring to figure 8.3: the links in pathway $\langle n_1, n_2, n_3 \rangle$ are compiled into a link between n_1 and n_3. Either there is no higher level of description than the one depicted in figure 8.3, or at the higher level n_1 and n_3 are not focus nodes. This allows for n_3 to be activated on the basis of the confirmation of n_1. For example, severe salmonellosis causes dehydration sufficient to cause hypotension; there is a compiled link between salmonellosis and hypotension.

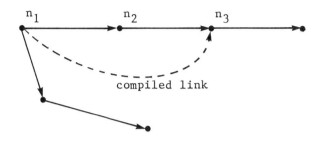

Fig. 8.3: Compiled links.

8.1.2 The Meaning of the Causal Link

ABEL gives a new interpretation to the causal relation from the usually coded interpretation (in the context of Expert Systems) of being "the likelihood of observing the cause/effect given the realization of the effect/cause". In ABEL the form of presentation of an effect and the likelihood of observing it depend upon various aspects of the presentation of the cause instance such as severity and duration, as well as on other factors in the context in which the causal phenomenon is manifested (such as the patient's age, sex and weight, and the current hypothesis about the patient's illness). Thus a causal link is interpreted to specify a **multivariate** relation between various aspects of the cause and the effect, taking into account the context and the assumptions under which the causal link is being instantiated (see figure 8.4).

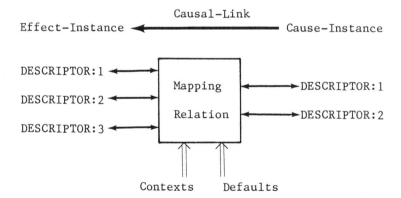

Fig. 8.4: Schematic description of a causal link
(adapted from Patil **et al**, 1982a).

Below we give a rule that describes the causal relation between diarrhea and metabolic-acidosis.

```
IF DIARRHEA    WITH (descriptor values)
               SEVERITY = "severe"
               DURATION = "more than 2 days"

THEN maybe     METABOLIC ACIDOSIS
               with (descriptor values)
               ANION-GAP := "normal"

                            ⎧ "mild"
                            ⎪ IF RECENT-THERAPY = "bicarb´ therapy"
               SEVERITY := ⎨
                            ⎪ "moderately severe"
                            ⎩ IF RECENT-THERAPY = "none"
```

For each causal link, therefore, the knowledge-base must specify the information characterizing that link, i.e. the extention of the multivariate relation.

Thus causal links are considered to be objects in their own right rather than simply as ordered pairs of nodes. This gives the system the capability of hypothesising the presence or absence of a causal link between two realized nodes. For example, "the patient suffers from severe metabolic-acidosis" can not be attributed to "the patient suffers from mild diarrhea" and thus the hypothesis of a causal link between metabolic-acidosis and diarrhea, is highly unlikely.

An important advantage of considering causal links as multivariate relations is that we can dynamically combine into one effect the separate effects of multiple causes (see below).

8.2 DYNAMICS

The dynamics of the system generate, manipulate and tie togeher the different levels of the knowledge-base into coherent patient-specific models (see Patil **et al**, 1981). There are operations for combining, into a single effect, the separate effects of multiple causal relations, and for decomposing effects into multiple causal relations. Referring to figure 8.5(a), suppose that metabolic-acidosis has been concluded and that diarrhea and shock have been observed. Attempting to attribute the entire metabolic-acidosis to either diarrhea or shock alone fails (i.e. the relevant cause-effect relations mismatch). However, combining the separate effects of the two causes, and thus partly attributing the metabolic-acidosis to diarrhea and partly to shock, yields a match.

An important case of multiple causes contributing jointly to a single effect arises whenever feedback is modelled; in any feedback loop there is at least one node acted on both by an outside factor and by the feedback loop itself (see figure 8.5 (b)).

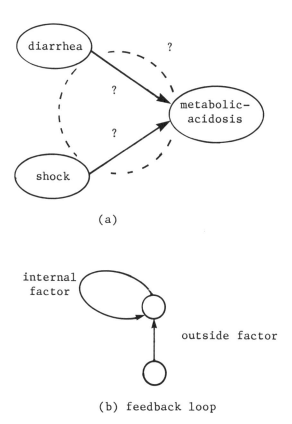

(a)

(b) feedback loop

Fig. 8.5: Combining multiple causes.

Other operations, summarize a description of a patient's illness at a given level into a description at the next more aggregate level; disaggregate a description at a given level into a description of the more detailed level; explore diagnostic possibilities by projecting backwards and fowards along a level of description.

8.2.1 Outlining the Diagnostic Strategy

Based on an initial set of serum electrolyte data and any other supplied findings, ABEL generates a small set of competing hypotheses each of which is a possible explanation of the electrolyte data. A hypothesis is an instantiation, at every level of detail, of a portion of the causal network. Each of these causal hypotheses or Patient Specific Models (PSMs) is **coherent** meaning that all the nodes involved in it are causally related and thus mutually complementary in their ability to explain the patient's condition. The competing PSMs can be compared at any level of detail.

A PSM corresponds to some interpretation of known facts. Complete pictures of the patient's illness that fit the interpretation can be generated by projecting causal chains in the PSM backwards to reach possible ultimate etiologies and then forward projecting uncompleted causal chains, either from the PSM or from the newly entered ultimate etiologies, and thus predicting new observable findings. This extended picture is known as the **diagnostic closure** of the PSM (see figure 8.6). Thus a diagnostic closure contains alternative extentions needed to complete, adequately, the explanation provided by the associated PSM.

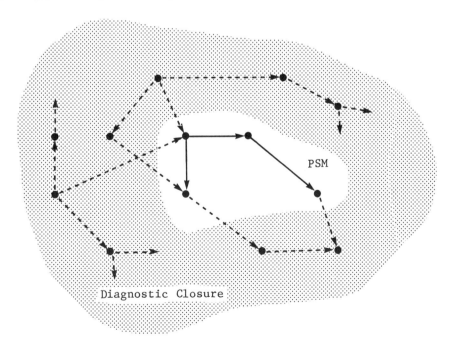

Fig. 8.6: Projecting a PSM to yield complete explanations of the patient's illness.

The promise of each PSM is evaluated and the diagnostic closure of the sufficiently promising PSMs are generated. Next, depending on the configuration of the PSMs' space (i.e. their number and relative degrees of promise), a carefully planned sequence of questions is generated and executed. Any new finding is taken into consideration and may generate new PSMs. The cycle of evaluating the promise of the competing PSMs, generating diagnostic closures, executing a sequence of questions, and updating the PSMs' space, is repeated until either some PSM provides a complete (every node is accounted) and coherent account for the patient's illness or there are no diagnostic closures for any of the PSMs (that is an exhaustive search of the PSMs' space has been made). The designers of ABEL, however, envisage the future incorporation of a knowledge intensive criterion for deciding when a satisfactory diagnosis has been reached (and the program should move onto therapy). This knowledge intensive criterion should weigh the costs of gathering further information in terms of morbidity, discomfort, time and money vs the benefits of better diagnosis in terms of an improved management plan and a more reliable prognosis. For example, in situations in which the management plan for each of the diagnostic possibilities is the same, further attempts to distinguish between alternatives do not have any utility. The decision to terminate the diagnostic process and move onto the therapy will be carried out by the global decision making component of the larger system.

The designers of ABEL believe that the consideration of a PSM as a hypothesis for a patient's illness, resulting from multiple diseases, is a more efficient, and a less error prone, way of building the complete picture of the illness than the ways adopted by systems like INTERNIST-I and PIP. These other systems consider each possible individual disease as a complete hypothesis. They attempt to build the complete picture of the patient's illness by incrementally piecing together interdependent components of a multisystem illness through promoting the consideration of diseases related to previously concluded/activated diseases.

Initial context formation

Below we describe, briefly, how the initial set of PSMs (causal hypotheses) is generated. This initial stage, which is very domain specific, is critical to the overall system performance.

Initially the program is provided with a set of serum electrolyte values and any other findings the user considers significant. The instances of the electrolyte data are present at the pathophysiological level. The set of possible competing single and multiple acid-base disturbances present in the patient

is generated from 'the acid-base nomograph' (see Patil,1981).
These provide competing explanations of the serum electrolyte
values.   The nodes at the clinical level, corresponding to these
disturbances, are instantiated; the instantiation of a node
includes determining the values of the node's severity, temporal
characteristics and other aspects relevant to the node.   Each
instantiated node at the clinical level is focally elaborated to
the pathophysiological level, i.e. the chain of focal links
linking it to some node in the pathophysiological level is
traversed, and the node at the end of the chain is instantiated.
Thus, portions of the pathophysiological network are instantiated,
i.e. the possible pathophysiological level explanations of the
electrolyte abnormalities, each explanation centering around one
of the likely acid-base disturbances, are generated.   After each
pathophysiological level explanation is completed it is aggregated
one level at a time to the clinical level of detail and this
completes the generation of one of the initial PSMs.   The set of
competing PSMs at every stage during a consultation are
represented with a single locus in terms of a tree structure,
referred to as the Patient Specific Data Structure (PSD) tree,
that allows different PSMs to share structure common between them.
Each node inherits from its superiors all the structure present in
them except that which is explicitly deleted (thus all the
observations are held on the topmost tree node).   The structure
visible from each leaf node of the PSD tree corresponds to an
individual PSM (see Patil, 1981).

8.2.2 Focusing Mechanism

One might say that ABEL's most important focusing strategy
is the use of the acid-base nomograph to home in on the problem as
quickly as possible.   However, as mentioned above this initial
priming facility can only be applied to the domain of acid-base
and electrolyte disorders.

In subsequent stages, in addition to the focusing potential
inherent in the focus nodes and links and the compiled links, the
diagnosis is focused by evaluating the promise of

(i)   PSMs, i.e. causal coherent hypotheses;
(ii) disease hypotheses contained in the diagnostic closure
      of some PSM.

Before we characterize ABEL's measures of promise, we will
first raise a distinction between a measure of belief or
confidence and a measure of promise.   Crudely, a measure of
promise is a measure of the 'distance' and the 'nature of the
journey' between what we have now (a hypothesis) and what we

require (an assertion); the most promising hypothesis is the one with the greatest potential of being asserted. Belief or confidence on the other hand, is a measure of the strength of confidence in our identification of what it is we think we have.

The designers of ABEL believe that those two measures should be distinct. They stress that a hypothesis could have a low belief measure and yet a high promise measure. They have an approach similar to CASNET's which employs the confidence factor as a belief measure and the weight as a promise measure of state hypotheses. However, they further indicate that they believe that the use of some threshold value on a measure as a confirmation criterion is arbitrary and thus difficult to explain. Hence in future developments of the system, the decision criteria as to whether a disease hypothesis or a PSM should be deemed to be confirmed or not (confirming a PSM yields a diagnosis termination criterion) will be part of the function of the system's global decision making component.

Evaluating the promise of a PSM

The promise of a PSM indicates how 'far' the PSM is from a complete explanation of the patient's illness. The promise of the ith PSM, P_i is given by:

$$P_i = \sum_j s_j$$

where: j ranges over the fully or partially unaccounted nodes in the PSM, and
s_j is the severity of the jth node.

A fully unaccounted node is one that is not an ultimate etiology and none of its potential causes are included in the PSM; a partialy unaccounted node is one that is only partially accounted by the presence of some of its causes in the PSM.

The above simple function for evaluating the promise of a PSM is intended to be modified in the future in order to:

1)   Take into consideration the need of a finding to be accounted for by an acceptable diagnosis; this would be analogous to the use of the global importance measures assigned to the manifestations in the INTERNIST-I system.

2)   Take into consideration the degree of explainability of a PSM; this is taken to be the minimal number of ultimate etiologies that collectively explain (cover for) all the unaccounted findings in the PSM.

Evaluating the promise of a disease hypothesis

ABEL's measure of the promise of a disease hypothesis, within the diagnostic closure of a promising PSM, takes into account the match of the reported findings expected on the hypothesis and also the 'covering power' of the hypothesis. Where 'covering power' is the hypothesis' capacity to explain findings presently unaccounted for; the procedure is as follows:

Hypotheses are grouped according to the number of unaccounted findings in the PSM that can be accounted for on each hypothesis. Within each group the hypotheses are assigned -- to borrow PIP's terminology -- a 'matching score' computed from the three factors:

1) **match:**   the number of causes and findings in the PSM that are consistent with the disease hypothesis.

2) **mismatch:**   the number of causes and findings in the PSM that are inconsistent with the disease hypothesis.

3) **unknown:**   the number of **unobserved** findings predicted by the hypothesis which are not inconsistent with the PSM.

Within each group the most promising disease hypothesis is the one that yields the highest score. Hypotheses that have a preponderance of findings in the PSM which are inconsistent with them (i.e. if match - mismatch $<$ Θ for some preset threshold value Θ) are discarded from consideration in the current diagnostic cycle. Hypotheses that insufficiently cover the findings in the PSM (i.e. if match + unknown $<$ Φ for another preset threshold value Φ) are completely removed from consideration.

8.2.3 Information Acquisition Mechanism

Every diagnostic cycle involves the execution of a carefully constructed plan of information gathering, the aim being to achieve a clinically meaningful and focused pursuit of diagnosis. The plan is in the form of a goal tree, the top level goal being to discriminate between the alternative explanations provided by the current set of PSMs and each leaf goal being to elicit some item of information (Patil **et al**, 1982b).

The global information acquisition strategy is to elicit information with the objective of discriminating between the competing possible solutions of some diagnostic problem. This general strategy has a number of specific derivatives depending on how the discrimination is carried out. The specific derivative selected for some problem depends on the configuration of its

possible solutions' space, i.e. the number of competing solutions
and their relative degrees of likelihood (promises). If the
discrimination is carried out by eliciting information concerning
the differences between the competing solutions then we have the
specific information acquisition strategy of **differentiation.** If
the discrimination is carried out by eliciting information that
favours/disfavours the most/least likely competing solution then
we have the specific information acquisition strategy of
confirmation/ruling-out. If the number of competing solutions is
large, and their relative degrees of likelihood are more or less
the same, a good policy is firstly to somehow **group** these
competitors and then apply a discrimination strategy of differentiation on the resulting groups. If the diagnostic problem in
hand is to determine which of the subentities of some concluded
class of entities is present (i.e. its possible solutions' set is
the set of subentities), then the discrimination strategy selected
is further classified as one of **refining.** Finally, if the
possible solutions' set is not clearly defined the discrimination
strategy is classified as one of **exploration.**

In ABEL the top level diagnostic problem is to determine the
complete picture of the patient's illness. The possible solution
set to this problem is the set of PSMs which are currently
promising. Depending on the configuration of this set, i.e. the
number of competing PSMs and their relative promises a discrimination strategy of **differentiate, confirm,** etc. is selected. Let
us suppose that a discrimination strategy of confirmation is
selected, i.e. the system decides to elicit information concerning
the most promising PSM. Thus the top level diagnostic problem is
reduced to the simpler problem of determining which of the
alternative complete explanations contained in the diagnostic
closure of the most promising PSM is the true account of the
patient's illness. If the possible solution set to this problem
is not clearly defined, i.e. the alternative explanations are
implicit, rather than explicit, in the definition of the
diagnostic closure, a discrimination strategy of **exploration** is
selected. This decomposes the top level diagnostic problem into a
set of simpler diagnostic problems, each of which needs to be
solved. One such simpler problem could be to determine the cause
of some unaccounted finding contained in the most promising PSM.
The possible solutions' set to this problem is the group of
disease hypotheses included in the diagnostic closure of the PSM,
each of which could separately account for the finding.

The discrimination strategy chosen depends on the number of
these hypotheses and their relative promises. Suppose that a
strategy of differentiation is selected. This means that the
system is going to concentrate on the differing aspects predicted
by the hypothesised diseases, i.e. elicit information about those

aspects. For example, if salmonellosis and acute renal failure are hypothesised as diseases likely to cause the acid-base discorder of metabolic-acidosis, then the patient's state of hydration is a good differentiation aspect to explore, since salmonellosis predicts dehydration (volume depletion) whilst acute renal failure predicts edema (fluid retention). Thus a diagnostic subproblem of the problem of differentiating between the causes of metabolic-acidosis is to determine the hydration state of patient.

This process of problem decomposition continues until the subproblems reached are ones that can be solved directly by asking single questions. The diagnostic information acquisition plan is thus completed.

The execution of a diagnostic plan results in the acquisition of new items of information from the user; this newly elicited information is used to update the collection of PSMs and thus the possible solutions' set of the top level diagnostic problem and a new diagnostic cycle begins, i.e. the generation of a new information acquisition plan begins. During the consultation the user can interrupt the diagnostic process to supply any information he/she wishes. If this information is in direct contradiction to the system's expectations about possible answers to a question, then before that information is incorporated in the PSMs, validity checks are carried out.

At this point it is interesting to compare ABEL's information acquisition mechanism with the corresponding mechanisms employed by its precedessor systems PIP and INTERNIST-I. INTERNIST-I has more or less the same information acquisition strategies (rule-out, pursue, discriminate, etc.); however these strategies are not employed to construct an orderly and focused plan of information acquisition, but to determine the item of information next to elicit from the user. The problem solvers of PIP and INTERNIST-I, alternate between gathering a fact (based on their hypothesis lists) and re-evaluating the hypothesis lists (based on the new fact). Each fact is treated as an independent inquiry.

8.3 FACILITIES

8.3.1 Explanatory Facilities

The goal tree encodes the program's rationale for asking a question; it explicitly encodes the program's reason for asking the question, its expectations about the information being sought, and how this information relates to the hypotheses under

(a) Goal tree.

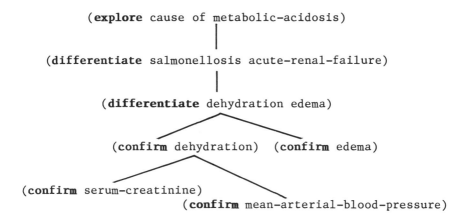

(b) **Explanation of the system's need to know the patient's status of serum creatinine.**

I am expecting the patient to have a mild elevation in serum creatinine. Increase in serum creatinine may be caused by dehydration, which may be caused by salmonellosis. The salmonellosis may account for the observed metabolic-acidosis. It is also the leading cause of metabolic-acidosis under consideration. Increase in serum creatinine may also be caused by acute renal failure, which may cause metabolic-acidosis.

(c) **Justification of the diagnostic reasoning behind the determination of the cause of the acid-base disorder metabolic-acidosis.**

I am exploring the cause of metabolic-acidosis. I am differentiating between the two leading causes of metabolic-acidosis, namely salmonellosis and acute renal failure. I am differentiating between dehydration and edema. The dehydration may be caused by salmonellosis and the edema by acute renal failure. I am pursuing dehydration. I am pursuing serum creatinine. Increase in serum creatinine may be caused by dehydration. Increase in serum creatinine may be caused by acute renal failure.

Fig. 8.7: Explanatory facilities
(adapted from Patil (1981), pp. 92-93).

consideration (these hypotheses define the context in which the question is being asked). ABEL is in a position to **explain** the usefulness of a question raised during a consultation and to **justify** (explain) its diagnostic reasoning. The explanation of the usefulness of an item of information enquired about, and the explanation of the system's diagnostic reasoning are, respectively, interpreted as ascend and descend of the goal tree. Figure 8.8 gives a goal tree and explanations generated from it.

The program is able to explain, by providing the English translation of any given level of the (leading) PSM, its under-standing of the patient's illness at various levels of detail.

8.4 IMPLEMENTATION DETAILS

The ABEL system uses XLMS (eXperimental Linguistic Memory System) to represent and manage its knowledge-base. The XLMS is an extention of LISP developed primarily by Hawkinson, Martin, Szolovits and others at MIT (Hawkinson, 1980). The English expl nation generator is implemented using the methodology developed by Swartout (1981).

REFERENCES

Hawkinson L.B. (1980): "X LMS: A linguistic memory system", MIT Laboratory for Computer Science, Technical Memo **MIT/LCS/Tm-173.**

Patil R.S. (1981): "Causal representation of patient illness for electrolyte and acid-base diagnosis", **MIT/LCS/TR-267.**

Patil R.S., Szolovits P. and **Schwartz W.B.** (1981): "Causal understanding of patient illness in medical diagnosis", Proc. **IJCAI-81,** pp. 893-899.

Patil R.S., Szolovits P. and **Schwartz W.B.** (1982a): "Modelling knowledge of the patient in acid-base and electrolyte disorders", in P. Szolovits (ed.), **Artificial Intelligence in Medicine,** AAAS Selected Symposium Series, West View Press, 1982, pp. 191-226.

Patil R.S., Szolovits P. and **Schwartz W.B.** (1982b): "Information acquisition in diagnosis", Proc. **AAA1-82,** pp. 345-348.

Swartout W.R. (1981): "Producing explanations and justifications of expert consulting programs", MIT Laboratory for Computer Science, Technical Report **MIT/LCS/TR-251.**

Chapter 9
NEOMYCIN

Application area:	Medicine.
Principal researchers:	W.J. Clancey and R. Letsinger (Stanford University).
Function:	To explicitly represent strategic knowledge and thus provide an efficient basis for teaching diagnostic reasoning, and interpreting student behaviour.

OVERVIEW

MYCIN's task specific knowledge is encoded in 'independent' production rules; this encoding does not make the knowledge involved perspicuous. It is not easy to see how the knowledge interrelates and how the rules interplay to implement the system's diagnostic and therapeutic tasks. The tutorial program GUIDON, which was associated with MYCIN, was unable to explain or properly evaluate students diagnostic guesses; i.e. it was unable to provide reasons for taking one measure next rather than another, and was unable to understand students shift of focus (Clancey, 1979). This is precisely because MYCIN uses a diagnostic model of reasoning that is different from how the domain experts reason. The NEOMYCIN system (Clancey and Letsinger, 1981 and Clancey, 1983a) is the outcome of an attempt to capture, more accurately, the expert's underlying competence and the conversational context.

Consider the following MYCIN rule:

RULE 543

IF: 1) the infection is meningitis
2) the subtype of meningitis is bacterial
3) only circumstantial evidence is available
4) the patient is at least 17 years old and
5) the patient is an alcoholic
THEN: there is suggestive evidence that diplococcus-pneumoniae is an organism causing the meningitis

The conditions in the rule antecedent serve different purposes. The first three conditions define the context of the rule application: Conditions (1) and (2) are present because the relevant etiological taxonomy can not be explicitly represented within the rule scheme; a top-down refinement strategy is, therefore, implicit in the ordering of such antecedent conditions. Condition (3) establishes the absence of hard (laboratory) evidence. In the presence of hard evidence a companion rule to the above allows the circumstantial evidence of alcoholism to be considered but gives it less weight. This principle of considering circumstantial evidence even when there are hard, physical, observations of the cause is not explicitly known to MYCIN. Condition (4) is known as a 'screening' condition; in the above case it prevents the system from asking if a child is an alcoholic. The presence of screening conditions in rule antecedents is necessary because commonsense knowledge ("world facts") are not explicitly represented in the knowledge-base, and thus the principle "establish the appropriateness of a question before putting it forward to the user" is only implicit in the system. Therefore, reasoning principles are implicit in the ordering of antecedent conditions. The authors of rules do not write them independently of their understanding of other rules. Because, for a rule to be usable by the system it must make a conclusion about a goal that is a condition of some other rule's antecedent, and its antecedent must be provided with means to evaluate its conditions -- other rules must be written and/or information must be sought from the user. The rule author composes the rule-base on some understanding of the whole but this understanding is left implicit in the rule-base.

In the vast majority of MYCIN rules only one antecedent condition remains once the contextual and screening conditions have been removed. In the example rule what we have is that "alcoholism suggests that diplococcus-pneumoniae is an organism causing the meningitis". The justification of this rule is, therefore, given in terms of the underlying causal mechanism that explains the association of the observation of "alcoholism" to the hypothesis "diplococcus-pneumoniae is an organism causing the meningitis". The justification of MYCIN rules, however, lie outside of the knowledge-base, i.e. the record of inference steps that ties antecedents to consequents has been left out. This is why rules very often are referred to as 'compiled knowledge'. They are compiled in the sense that they are evolved patterns of reasoning that cope with the demands of ordinary problems and thus leave out "unnecessary" steps. The penalty paid, from this omission, is a reduction in the system's flexibility as a problem solver. Human experts derive their flexibility as problem solvers from their ability to **violate** a rule in difficult, "non-standard" situations, by being able to reason with the knowledge involved in

the inference steps that tie antecedents to consequents. These
rule justifications might not be necessary for the handling of
standard cases, but their presence is necessary for the tutoring
of neophyte physicians (refer to Clancey (1983b) for a detailed
exposition of the weaknesses of the MYCIN framework).

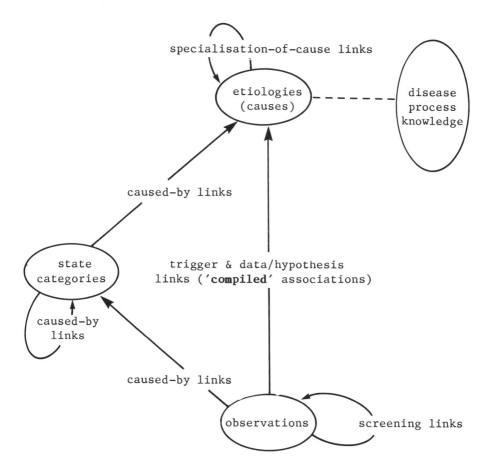

Fig: 9.1: Illustrating the conceptual structure of
NEOMYCIN's factual knowledge.

9.1 STATICS

The NEOMYCIN system separates and makes explicit the factural
(declarative) knowledge, the reasoning (procedural) knowledge and
the relationships between the two. The reasoning knowledge is
abstracted (Clancy, 1983c), i.e. it is expressed in domain-

independent terms, where 'domain-independence' is relative to a
constrained class of domains. The relationships between reasoning
knowledge and factual knowledge indicate, for each component of
the reasoning knowledge, the structural component/s of the factual
knowledge it draws from, and thus make the reasoning knowledge
concrete.

9.1.1 Factual Knowledge

Below we explain the basic structure of NEOMYCIN's factual
knowledge. This is depicted in figure 9.1.

Etiologies

The diseases explaining the patient's illnesses (etiologies)
are classified in terms of an **etiological taxonomy**, a portion of
which is given in figure 9.2. The representation of a disease/
disease category, as a process, is given in terms of a frame

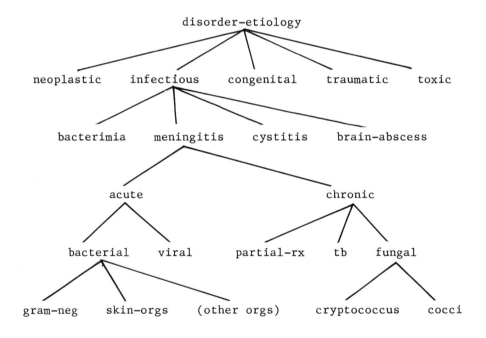

KEY
——————— : specialisation-of-cause link

Fig. 9.2: Portion of etiological taxonomy
(adapted from Clancey and Letsinger, 1981).

associated to the relevant node of the etiological taxonomy; slots
are process descriptors such as "extent", "location", "course"
(i.e. progression of symptoms) etc. Also specified are follow-up
(pinning down) questions that should be asked immediately when the
disease is implicated (e.g. to establish when symptoms occurred
and their ordering and change in severity). This disease process
knowledge is considered orthogonal to the etiological taxonomy.

Observations

Observations (findings) are divided into soft and hard
findings; the former represent circumstantial or historical items
of information, and the latter represent items of information
obtained through laboratory tests or direct measurements. The
characterizing features of findings are also specified; the
observation of a finding prompts the asking of questions for
eliciting values for the entire set of features associated with
the finding. There are relationships between observations
denoting abstractions and restrictions on them, e.g. "if a patient
is not immunosuppressed, then he is not an alcoholic". Such world
facts (commonsense knowledge) are represented in terms of
'screening rules' which chain together to form a hierarchy (see
section 9.2.2).

Observations represent evidence for hypothesised etiologies.
Associations between findings and etiologies are represented in
terms of two types of rules:

i)   **trigger rules:** these rules associate combinations of
     findings to etiologies when the presence of the latter
     is strongly suggested by the presence of the former,
     e.g. "stiff neck and headache is strongly suggestive of
     meningitis". Tigger rules are used for generating new
     hypotheses.

ii) **data/hypothesis rules:** like the trigger rules, the data/
     hypothesis rules associate combinations of findings to
     etiologies; however, unlike trigger associations, such
     associations are not sufficiently suggestive, on their
     own, to generate new hypotheses, but they become of
     importance in the context of the associated or related
     hypotheses.

Trigger and data/hypothesis rules are examples of 'compiled'
associations, subsuming underlying causal chains; these chains
constitute the justification of these rules and are reduntantly
stored in the knowledge-base, as canned text, for explanation
purposes.

Causal network

The causal network links observations to etiologies via state categories (pathophysiological states or categories of disease). The network is implicitly defined through a colleciton of **causal rules** representing "caused-by" associations of specified degrees of strength.

9.1.2 Reasoning Knowledge

The overall diagnostic reasoning knowledge is functionally decomposed to yield a hierarchy of **reasoning tasks** representing important functions and subfunctions of the diagnostic process (see figure 9.3). Examples of such tasks are: initial context formation (identify problem), history taking subtask (establish hypothesis space), etc. In figure 9.3 non-terminal nodes stand

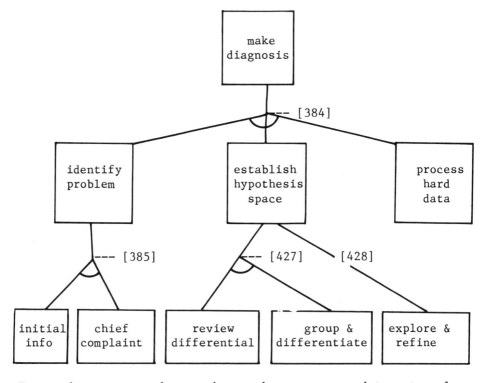

KEY: nodes correspond to tasks; numbers correspond to meta-rules.

Fig. 9.3: NEOMYCIN'S static representation of its diagnostic reasoning knowledge (adapted from Clancey and Letsinger, 1981).

Table 9.1

Task	Meta-rule/s
Group-and-Differentiate	if: there are two items on the differential (working memory of hypotheses) that differ in some disease process feature then: ask a question that differentiates between these two kinds of processes
Explore-and-Refine	if: the hypothesis being focused upon has a refinment that has not been pursued then: pursue that refinement
Test-Hypothesis	if: the datum in question is strongly associated with the current focus then: apply the related list of rules if: the datum in question makes the current focus more likely then: apply the related list of rules
Characterize-Data	if: there is a datum that can be requested that is a characterizing feature of the recent finding that is currently being considered then: findout about the datum
Findout	if: the desired finding is a subtype of a class of findings and the class of findings is not present in this case then: conclude that the desired finding is not present

for tasks that invoke other tasks while terminal nodes stand for tasks that manipulate the domain factual knowledge and ask questions. A reasoning task is statically represented in terms of a set of **meta-rules** that collectively constitute a procedure for achieving the task. A meta-rule has a premise which indicates when the meta-rule is applicable and an action, indicating what should be done whenever the premise is satisfied. Premises refer to the current list of hypotheses, the recent observations reported, the causal network, the etiological taxonomy or the disease process knowledge. Actions could be to apply domain (object) rules which elicit additional observations or which invoke other tasks. Meta-rules are expressed in abstract and not concrete terms, i.e. they refer to domain-independent concepts like "findings" and "causes" rather than domain-dependent concepts like "shaking chills" and "infection". Table 9.1 gives some specific examples of tasks and their associated meta-rules. Currently the diagnostic reasoning is statically represented in terms of 30 tasks and 74 meta-rules.

The components of the factual knowledge used by some reasoning task are those components manipulated by the terminal tasks subsumed under the given task. The uses of a factual knowledge component are obtained by tracing the various tasks that use it.

9.2 DYNAMICS

The meta-rules representing some task are treated as a pure production system (Davis and King, 1977). The rules are repeatedly tried, in order, returning to the head of the list when one

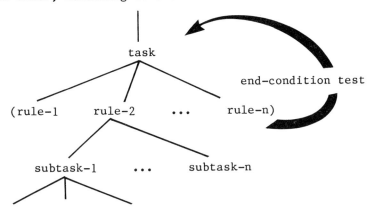

Fig. 9.4 : Rule-based invocation and interruption of reasoning tasks (adapted from Clancey and Letsinger, 1981).

succeeds. Thus the same subtask can be invoked multiple times by
a given task, stopping (i.e. returning control to the supertask
that invoked the given task) when no rule succeeds, or an end of
task condition is true. The end of task condition is itself
determined by rules (see figure 9.4); the termination of a task
implies the termination of all the subtasks subsumed under it.
The top level task of "making a diagnosis" is automatically
invoked at the beginning of a consultation.

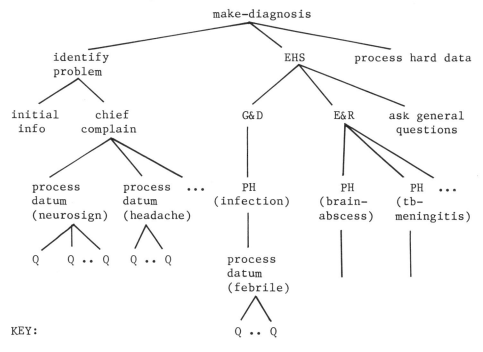

KEY:

EHS: Establish-Hypothesis-Space; G&D: Group & Differentiate;
E&R: Explore & Refine; PH: Pursue-Hypothesis; Q: Question.

Fig. 9.5: Dynamic generation of diagnostic plan.

Figure 9.5 illustrates a portion of the diagnostic plan,
dealing with some problem case, that is dynamically generated by
instantiating reasoning tasks.

Observations constitute evidence for hypothesised etiologies
and state categories. The set of active hypotheses is referred to
as the **'differential'**. General causes belonging to the
differential are subsequently replaced by their more specific
descendants and state categories by their prior causes and

ultimately by diagnostic hypotheses in the etiological taxonomy.
As was mentioned earlier, evidence is classified as soft
(circumstantial) or hard (physical observations). The reasoning
behind the combination of hard and soft evidence is made explicit
in the system. This basically involves the dynamic readjustment
of the support provided to hypotheses by soft evidence in the
presence and absence of hard evidence.

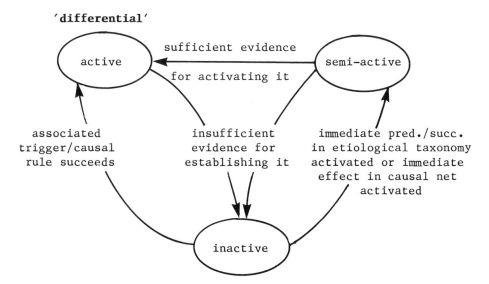

'**differential**'

Fig: 9.6: Basic hypothesis status transitions.

9.2.1 Outlining the Overall Diagnostic Strategy

The initial stage of the diagnostic process is to identify
the problem at hand, i.e. to formulate the initial context from
which the subsequent investigation would be directed. This
preliminary stage involves the elicitation from the physician user
of the system, of standard initial information concerning the age,
sex and race of the patient followed by the chief complaint/s.
These items of information are used by the system to determine
whether any trigger rules succeed; if some trigger rule requires
more information before it can be applied, then the system asks
about the relevant information (e.g. the observation of headache
prompts the asking of whether the patient has a stiff neck as well
in order to make the trigger rule for meningitis applicable). The
etiologies suggested by the successful trigger rules become **active**
hypotheses, i.e. elements of the differential. State categories
strongly suggested by this initial information, via causal rules,
are also included in the differential. Immediate

successors/predecessors of hypothesised etiologies as well as
prior causes of hypothesised state categories could be considered
as **semi-active** hypotheses, i.e. hypotheses focused by the
differential rather than by conducted observations. Such semi-
active hypotheses are activated (possibly replacing, in the
process, their related hypotheses from the differential) when
sufficient evidential support towards them is obtained; otherwise
they are **inactivated**. Figure 9.6 depicts, at a high level of
abstraction, the NEOMYCIN diagnostic strategy.

Very crudely, subsequent stages of the investigation are
involved in the updating of the space of active and semi-active
hypotheses until the differential contains only the most specific
etiologies, i.e. etiologies corresponding to terminal nodes of the
etiological taxonomy. This requires alternating between
collecting and processing relevant soft and hard evidence.

9.2.2   Focusing and Information Acquisition   Strategies

The main focusing strategies in the NEOMYCIN system is the
use of the trigger rules to constrain the generation of new
hypotheses to only those that are strongly suggested by the
observations. In the case of broad domains, with large
etiological taxonomies, the use of the above focusing aid in
conjunction with the reasoning strategy "work from the general to
the specific" is necessary to focus the system into the middle of
the taxonomy.

Other focusing strategies determine which active/semi-active
hypothesis should next be considered by the system.

The diagnostic plan that is dynamically generated via the
instantiation of meta-rules (e.g. the plan in figure 9.5) is
basically a plan that directs the meaningful acquisition of
additional information from the user. The plan, therefore, makes
explicit the diagnostic rationale (strategic reasoning) behind the
asking of some question. More specifically, high level
information acquisition strategies are concerned with differentiating between competing hypotheses, or with the refining of
hypotheses, or with the pursuing of hypotheses, etc. whilst low
level information acquisition strategies are conncerned with fully
characterizing the supplied findings, or a hypothesised etiology
(pinning down questions) or the problem at hand (general questions
to determine the completeness of the case history). Another low
level information acquisition strategy uses the screening rules
(commonsense knowledge) to ensure that no unintelligent questions
are being made, e.g. preventing the asking of whether the patient
is alcoholic when it is already specified that the patient is not

immunosuppressed. Screening rules specify conditions that have to be met before a question can be appropriately asked. They are appied in a backward reasoning fashion in an attempt to deduce a required item of information from the information already supplied by the user, without asking the user.

9.3 FACILITIES

9.3.1 Explanatory Facilities

As we mentioned earlier, the diagnostic plan (strategy tree) makes explicit the strategic reasoning behind the asking of some question by the system, i.e. the context of the question is made explicit. For example when formulating the initial context of the problem, the system asks questions in order to make trigger rules applicable and when it wants to find out about an item of information it uses screening rules to prevent it from asking an unintelligible question. Furthermore, it asks questions in order to differentiate between hypotheses, test hypotheses, fully characterize observations, etc. Thus there are different reasons **why** a question is being asked by the system. Any satisfactory explanation as to the usefulness of the item of information which the question attempts to elicit, should, therefore, make explicit (reflect) the strategic reasoning that gave rise to the question.

NEOMYCIN's explanations are generated at the level of general (abstract) strategies, instantiated with factual domain knowledge, when possible, to make them concrete. Such explanations are referred to as abstract or strategic explanations. Below we outline NEOMYCIN's explanation system. A fuller account can be obtained from Hasling (1983) or Hasling **et al** (1984).

Strategic WHY and HOW explanations

A strategic WHY explanation refers to the usefulness of a task undertaken by the system. The terminal nodes of the strategy tree denote tasks that request items of information and apply object-rules. A task is invoked by a meta-rule in another task and thus a strategic WHY explanation for some task is given by stating the task and meta-rule that invoked the given task. Subsequent WHYs would therefore result in ascending the strategy tree.

A strategic HOW explanation refers to the achievement of a task undertaken by the sysem. The explanation presents any meta-rules associated with the relevant instance of the task that have

been completed, as well as the one currently being executed.
Subsequent HOWs would therefore result in descending the strategy
tree.

An interesting property of WHY and HOW explanations resulting
from the structure of tasks and meta-rules is that a WHY
explanation essentially states the premise of the meta-rule; this
is exactly the reason the meta-rule succeeded. A HOW explanation
is a statement of action of a meta-rule; this is exactly what was
done.

Besides the strategic WHYs and HOWs the user of the NEOMYCIN
system can ask about the current hypothesis, the set of hypotheses
currently being considered, and evidence for hypotheses at the
domain level.

The nearest to NEOMYCIN's strategic explanations are ABEL's
explanations; ABEL's goal tree is analogous to NEOMYCIN's strategy
tree. Using NEOMYCIN's terminology (for describing its reasoning
knowledge), a diagnostic task for ABEL is to discriminate the
competing solutions to some problem. The strategies of differentiate, confirm, rule-out, group-and-differentiate, refine and
explore collectively constitute a procedure for carrying out the
task and hence each one could be coded into a meta-rule, the
premise of which would give the justification for employing the
particular discrimination strategy in some context, and the action
of which would give the steps of the strategy (i.e. the steps for
achieving the discrimination which could be invoking other
discrimination strategies). In ABEL the usefulness of questions
raised by the system are explained in the context of the
particular discrimination strategies leading to them. Similarly,
its conclusions are explained in the context of the strategies
employed for establishing them. However, neither the
justifications for undertaking, nor the actions carried out by,
particular instances of discrimination strategies, are included in
such explanations.

9.3.2 Teaching Facilities

NEOMYCIN can be interfaced with GUIDON2 (Clancey, 1979), the
successor of the GUIDON program.

REFERENCES

Clancey W.J. (**1979**): "Tutoring rules for guiding a case method
dialogue", **Int. J. Man-Machine Studies, Vol. 11,** pp. 25-49.

Clancey W.J. and Letsinger R. (1981): "NEOMYCIN: Reconfiguring a rule-based expert system for application to teaching", Proc. **IJCAI-81**, pp. 829-836.

Clancey W.J. (1983a): "Methodology for building an intelligent tutoring system", in Kintsch, Polson and Miller (eds.), **Methods and Tactics in Cognitive Science,** Lawrence Erlbaum Publishers.

Clancey W.J. (1983b): "The epistemology of a rule-based expert system - a framework for explanation", **Artificial Intelligence, Vol. 20,** pp. 215-251.

Clancey W.J. (1983c): "The advantages of abstract control knowledge in expert system design", Proc. **AAAI-83,** pp. 74-78.

Davis R. and King J. (1977): "An overview of production systems", **Machine Intelligence, Vol.8,** pp.300-332.

Hasling D.W. (1983): "Abstract explanations of strategy in a diagnostic consultation system", Proc. **AAAI-83,** pp. 157-161.

Hasling D.W., Clancey W.J. and **Rennels G.** (1984): "Strategic explanations for a diagnostic consultation system", **Int. J. Man-Machine Studies, Vol. 20,** pp. 3-19.

Chapter 10
CRIB

Application area:	Fault diagnosis.
Principal researchers:	F.H. George (Brunel University), T.R. Addis (ICL - now at Brunel University), and R.T. Hartley (Brunel University - now at Kansas State University).
Function:	Diagnosis of faults in computer hardware.

The Computer Retrieval Incidence Bank (CRIB) system (George, 1977; Addis, 1980 and Hartley, 1981 & 1984) represents one of the early British attempts (carried out between 1974 and 1979) to automate the process of fault finding in machines using predominantly heuristic rather than the, then, standard algorithmic methods. The aim of CRIB was to closely resemble a human fault finder at his/her best. The three features that characterized CRIB from previous attempts in fault diagnosis were:

1. heuristics
2. anticipatory information
3. flexible interactive language

Although no version of CRIB is currently operational, a lot can be gained through an understanding of the analysis that went into the design of this first generation system. CRIB should be particularly notable for its learning aspect, i.e. its ability to adapt (revise) its 'knowledge' in the light of experience gained from investigations carried out both by the system and by engineers ('live' data). Very few systems of that period exhibited this feature. In CRIB's case, however, since the knowledge-base was to be initialised from the crude knowledge extracted from the available fault finding guides, this primitive form of learning was necessary in order to incrementally revise this knowledge.

What is to be described here refers to the initial version of CRIB. A subsequent, faster, version was based on ICL's Content

Addressable File Store (CAFS) MK I system (Addis and Hartley, 1979 and Addis, 1980). This version speeded up the diagnostic process by making the backtracking aspect of it redundant.

OVERVIEW

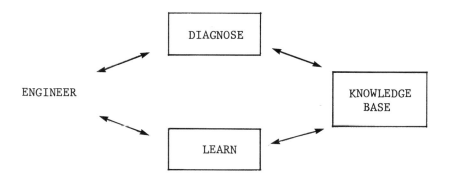

Fig. 10.1: Overview of the CRIB system.

In CRIB the 'patient' computer is modelled as a taxonomy of units whose leaf nodes represent replacable or repairable units. Each unit is associationally related to groups of symptoms. One group represents the "sufficient" condition and the others represent "necessary but not sufficient" conditions for establishing the presence of some fault in the unit.

These groups of symptoms are continuously revised in the light of new experience and thus they form the basis for a knowledge intensive mechanism for hypothesising faulty units from observed and assumed symptoms.

Diagnosis involves determining the 'route' through the unit taxonomy leading to the fault by matching groups of symptoms. In this respect, CRIB employs the domain independent diagnostic principle, "work from the general to the specific". This principle -- also employed in the INTERIST-I, CADUCEUS and NEOMYCIN systems -- functions to control the proliferation of active hypotheses. If the route under investigation is not promising, then the system backtracks one level in the taxonomy and attempts an alternative route. The designers of CRIB list a number of information acquisition heuristics for evaluating the usefulness of actions to be suggested next to the engineer user of the system. Actions function to acquire further symptom information with the aim of obtaining matches with groups of symptoms under investigation. CRIB's diagnostic reasoning is basically categorical.

CRIB is meant to be an aid to the engineers rather than their replacement; the system accepts observations conducted by the engineer and suggests actions to elicit further observations. The engineer, however, is not obliged to follow these actions. He/she is allowed to redirect the program to investigate a particular unit or subunit. The language used in the communication between the engineer and CRIB is a form of jargon English. (Figure 10.1 gives an overview of the CRIB system. Figure 10.2 gives the basic conceptual structure of the knowledge-base, of the CRIB system.)

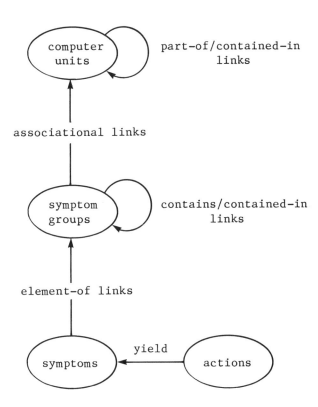

Fig. 10.2: Basic conceptual structure of CRIB's knowledge-base.

10.1 STATICS

The CRIB knowledge-base contains information on machine subunits, symptoms and their interrelationships, as well as information relating to actions to discover symptoms.

10.1.1 Model of the Computer

The 'patient' computer's structure is modelled as a taxonomy of hardware units and subunits, the terminal nodes of which represent those units that are either replaceable or repairable. A simple example of the first few levels of such a taxonomy is given in Figure 10.3.

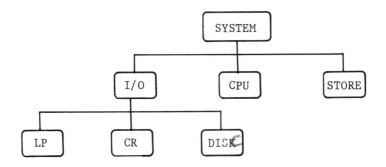

Fig. 10.3: A simple taxonomy of subunits of the 'patient' computer.

10.1.2 Symptom Groups

Each machine subunit is associated with:

i)   A **total group** (T-group) of symptoms, i.e. the accumulated group of symptoms that is observed to occur during the many investigations of faults related to the subunit.

ii) A **key group** (K-group) of symptoms. This is the subset of symptoms of the T-group, whose presence is sufficient to establish the presence of some fault related to the subunit.

iii) A set (possibly empty) of **subgroups** (S-groups) of symptoms. Each S-group is a subset of the K-group. Thus it represents a group of symptoms whose presence is necessary but not sufficient to establish the presence of some fault related to

the subunit. A success ratio is taken as a measure of the 'degree
of sufficiency' of an S-group and this is used to determine
whether or not it could be used by the diagnostic process to yield
matches with observed and assumed symptoms. (See section 10.2.3 on
learning.)

The CRIB knowledge-base was initialised using the available
fault finding guides for the ICL 2903 machine. It comprised fifty
five units forming a six-level taxonomy and three hundred and
seventy eight symptoms arranged in seven hundred and thirty eight
groups. The main difficulty encountered in so doing, was that
such groups were likely to fall into neither category of K-group
nor S-group. The solution was to specify each group both as a K-
group and as an S-group and to let the learning mechanism revise
this preliminary, rather crude, configuration of symptom groups.

10.1.3 Actions

An action functions to disclose a symptom or its inverse.
For example, the action 'run store test program X' results in the
symptom 'store works', or its "inverse" 'store faulty'. The CRIB
knowledge-base contains information on every action. This information gives a brief description of the action and relates it to
the symptoms concerned. Also specified are the approximate time
taken to perform the action, the level of training required for an
engineer to be able to perform it, and the number of times it had
so far been used in the investigation.

10.2 DYNAMICS

10.2.1 Outlining the Overall Diagnostic Strategy

Symptoms provide evidence for hypothesised faulty units. The
objective of CRIB'S diagnostic strategy is to determine, in the
shortest possible time, a route through the taxonomy to the fault.
At every stage the system would attempt to determine which subunit
(if any) of the currently most specific hypothesised-as-faulty-
unit, is most, and sufficiently, likely to be the faulty one. If
none of these subunits yields a sufficiently promising hypothesis
then, depending on how promising such potential hypotheses are,
the system would either backtrack to the superunit or attempt to
acquire additional symptom information (through the use of a
repertoire of powerful heuristics) in order to enhance the promise
of some potential hypothesis to the point of meeting one of the
"worth-pursuing" criteria. These are, in descending order of

preference: "best complete match on K-group", "best complete match
on S-group". The satisfaction of a worth-pursuing criterion
results in a drop in level and the next hypothesis to be pursued
is that the relevant subunit is faulty.

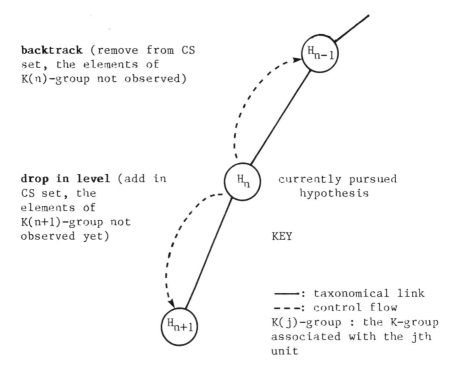

backtrack (remove from CS
set, the elements of
K(n)-group not observed)

drop in level (add in
CS set, the
elements of
K(n+1)-group not
observed yet)

currently pursued
hypothesis

KEY

———: taxonomical link
– – –: control flow
K(j)-group : the K-group
associated with the jth
unit

Fig. 10.4: Updating the CS set.

The initial hypothesis to be pursued is "the machine is
faulty". This hypothesis predicts (assumes) the symptoms belonging to the K-group associated with the topmost unit. The initial
set of symptoms observed by the engineer are entered either in
code form, e.g. S1036 or the inverse N1036, or in jargon English,
e.g. "store works". The set of observed and assumed symptoms is
called the Current Symptoms (CS) set. Every time there is a drop
in level, this set is updated to include any symptoms belonging to
the K-group of the relevant subunit that are not already included
in the CS set, and are not refuted through observing their
inverses. Similarly every time the system backtracks, the CS set
is updated by deleting any symptoms whose inclusion in the set was
solely based on predicitons made by the revoked hypothesis (see
figure 10.4). New symptoms observed by the engineer are entered
into the CS set.

Borrowing PIP's teminology for classifying its potential hypotheses, the CRIB hypotheses could be classified as **inactive, active,** and **semi-active.** Thus the CRIB diagnostic process could also be abstracted in terms of a hypothesis status transition diagram (see figure 10.5).

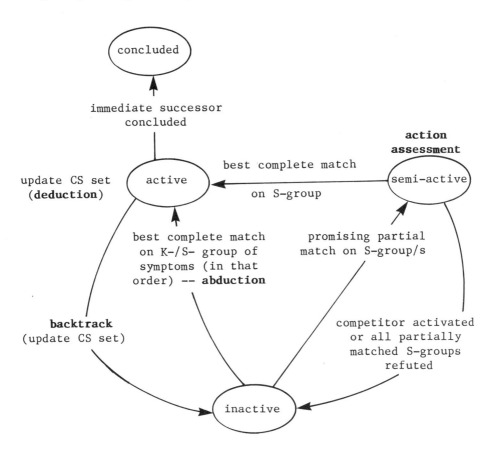

Fig. 10.5: Hypothesis status transition diagram.

The first hypothesis to be activated is that the machine is faulty. The question then arises as to which of the component units is faulty. The process of refinement continues through the taxonomy. If no successor hypotheses of a hypothesis meet any of the worth-pursuing criteria, then, if none of them have a sufficiently high promise to justify seeking to meet a worth-pursuing criterion, the hypothesis is inactivated. On the other hand, if one, or more, of them is sufficiently promising it/they are semi-activated. Actions are suggested to yield symptoms and obtain an S-group match on a semi-activated hypothesis. A semi-active hypothesis is acitivated if it has the best S-group match. A semi-active hypothesis is inactivated, either if a competing semi-active hypothesis is activated (by meeting a worth-pursuing criterion) or if none of its partially matched S-groups of symptoms could be completely matched. If all the semi-active hypotheses associated with an active hypothesis are refuted, then the active hypothesis is also refuted (inactivated).

If a leaf node of the unit taxonomy is reached, and the fault in question is confirmed by the engineer, then the hypothesis asserting the presence of that fault is concluded. The conclusion of a hypothesis causes its immediate predecessor to be concluded as well.

If, on the other hand, a leaf node is reached and the fault is not confirmed this means one of four things:

i)   a route taken, based upon a chosen S-group, did not lead to the fault;
ii)   a K-group is incomplete;
iii) the taxonomy is incapable of reflecting the fault location precisely enough;
iv)   there are multiple, transitory or 'soft' faults.

The system would then backtrack automatically to the last decision point and would try to find another match. If the whole taxonomy is searched and the fault is still present, then the system is incapable of reasoning about that fault.

10.2.2 Focusing Mechanism

At every stage during an investigation the scope of attention of the diagnostic process is dictated by the taxonomy of units. The focus is the successors of the currently pursued-as-faulty-unit.

The CRIB focusing mechanism evaluates the promise of some successor in terms of an estimate of the likelihood that it lies on the route to the fault. The promise of the hypothesis depends on:

1) **agreement:**   The degree to which the CS set covers the hypothesis' expectations. The special cases of importance of this criterion are: complete match on K-group, and complete match on an S-group.

2) **disagreement:** The degree to which the hypothesis disagrees

with the observed symptoms. This depends on the range of observed symptoms that did not belong to the T-group of the hypothesis. Such symptoms are referred to as **'unexpected'** since they are not deducible from the hypothesis. They could even refute symptoms expected on the hypothesis.

Thus the promise of a hypothesis depends on the nature of the match between the hypothesis' expectations and the CS set. Figure 10.6 (a)-(h) summarizes the eight possible classes of match that could arise between a hypothesis and the CS set, arranged in descending order of 'goodness'.

KEY (to figure 10.6)
  OS: Observed Symptoms;
  CS: Current Symptoms (those both observed and assumed on the previous hypotheses);
  Tg: entire set of symptoms expected on the hypothesis to be evaluated;
  Kg: set of symptoms which when observed are sufficient to establish the hypothesis;
  Sg: symptoms necessary on the hypothesis to be evaluated (subset of Kg);
    : unexpected symptoms.

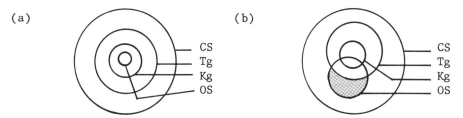

(a) Complete match on T-group (ideal case).
(b) Complete match on T-group with unexpected observations.

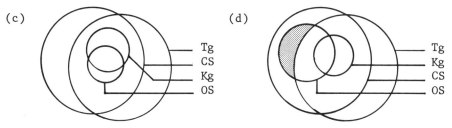

(c) Complete match on K-group without unexpected observations.
(d) Complete match on K-group with unexpected observations.

continued..

continuation..

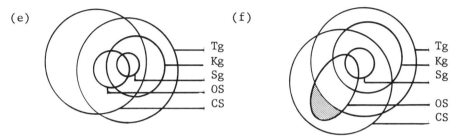

(e) Complete match on S-group without unexpected observations.
(f) Complete match on S-group with unexpected observations.

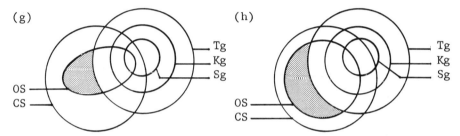

(g) Partial match on S-group with sufficiently small proportion
    of unexpected observations (promising partial match).
(h) Partial match on S-group with far too many unexpected
    observations (non promising partial match).

Fig. 10.6: Classes of match (adapted from George, 1977).

10.2.3 Information Acquisition Mechanism

When there is no complete match with any of the S-groups
associated with any of the successors, but some of these partial
matches are promising (i.e. they belong to the class of match
depicted in figure 10.6(e)) the diagnostic process calls upon the
services of the information acquisition mechanism. The function
of this mechanism is to determine the best actions to be performed
and to assess their outcomes. The objective behind it is to
obtain either a complete match with some S-group or a refutation
of every group. The usefulness of an action is based on the three
criteria:

1. Time to perform
2. Likelihood of completing an S-group
3. Likelihood of eliminating S-groups

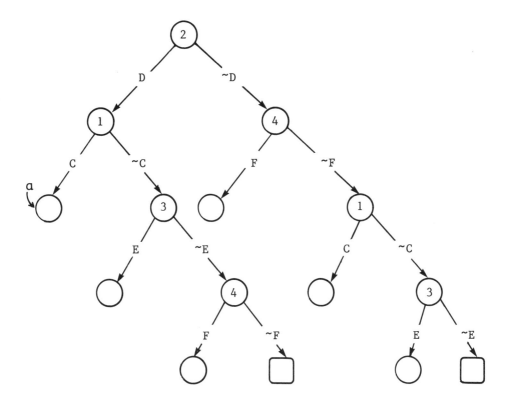

Fig. 10.7: Decision tree characterized by "most mentions unless a singleton S-group is better".

A heuristic that captures the above three criteria is: "Select the action that yields the most mentioned symptoms unless there is an S-group with one outstanding symptom only **and** the relevant action has a shorter total time of performance". Below we demonstrate the use of this heuristic (adapted from George, 1977).

Suppose that six S-groups are incomplete and they lack the following symptoms:

| S-group | 1 | C | 2 | C, D | 3 | D, E |
| | 4 | ~C, E | 5 | ~D, F | 6 | D, ~E, F |

The relevant actions are (assuming that the same action will give rise to either the symptom or its inverse with equal probability):

Action 1 to find C has a time to perform of 10 mins.
Action 2 to find D has a time to perform of 5 mins.
Action 3 to find E has a time to perform of 20 mins.
Action 4 to find F has a time to perform of 3 mins.

Figure 10.7 gives the decision tree obtained using the above heuristic. Each numbered circle represents the performance of an action. For each action there are two outcomes the observation of a symptom or its inverse -- hence two branches spring from each circle. A branch can terminate when either:

1)   a complete match with an S-group is obtained, e.g. node "a" where complete matches with S-groups 1 and 2 are obtained, or

2)   all S-groups are eliminated; such terminal nodes are pictorially represented in figure 10.7 as squares.

Thus according to the heuristic the best action to perform first is action 2: symptoms D and ~D are the most mentioned symptoms and although symptom C is the only outstanding symptom in S-group 1, its associated action, action 1, takes more time to perform than action 2. Suppose that the outcome of action 2 is "~D", then S-groups 2, 3 and 6 are eliminated and the outstanding symptoms in the remaining S-groups are: 1 C; 4 ~C,E; 5 F. The best action to carry out then is action 4, since its associated symptom, F, is the only outstanding symptom in S-group 5, and this action takes less time to perform than action 1 associated with the most mentioned symptoms C and ~C.

A well-known technique from graph theory was employed by the designers of CRIB to determine the average time to reach a definite conclusion, i.e. either a completed S-group or all S-groups eliminated: Each terminal node is assigned a "score" given by the time taken to perform the action which results in reaching the node. These scores are propagated upwards to give scores for non-terminal nodes as the average of the scores of their two successor nodes plus the time taken to perform the actions that results in reaching them. The score asigned to the top-most node gives the required result. For the example tree the average time to either complete an S-group or to refute all S-groups is around 20 mins.

Although CRIB's information acquisition mechanism is based on the above heuristic, the designers of the system list a repertoire of alternative heuristics: "shortest action", "shortest S-group", "most mentions" and "longest S-group". This indicates that the mechanism is so designed so as to be able to facilitate a change if it was thought necessary.

10.2.4 Learning Mechanism

The learning mechanism was considered by CRIB'S designers as the feature that made the system feasible. The original idea was of a mechanism that consists of two interacting levels of adaption of increasing complexity (see figure 10.8). The low level would be based on the principle of simple reinforcement, where successes are repeated and failures are eliminated. The high level would be based on the principle of selective reinforcement, where "succcess" reinforces positively and "failure" reinforces negatively. The function of the mechanism being to revise the system's "knowledge" in the light of experience gained from investigations carried out by the system and those carried out by the engineers who tackled problems of special interest or completed CRIB's unsuccessful investigations.

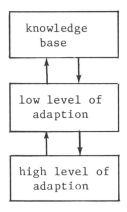

Fig 10.8: CRIB's learning mechanism.

The implemented version included only low level adaptation which was sufficient to prove the principles could be applied in a simple and effective manner.

Low level of adaption

The revisions to the system's "knowledge" brought about by this level of learning involves the taking up of new S-groups and the dropping out of old ones. Each S-group is assigned a success ratio, which can be interpreted as the degree to which the presence of the S-group supports the assertion that the relevant unit is faulty. The success ratio is given by the ratio of the number of times, N_S, the S-group was used to yield a drop in level en route to a fault, to the total number of times, N_T, it was used to yield a drop in level (not necessarily en route to a fault).

Referring to figure 10.9, S-groups SG2, SG3, SG7 and SG8 would have both their N_S and N_T values incremented by 1 whilst S-groups SG4 and SG5 have have only their N_T values incremented by 1.

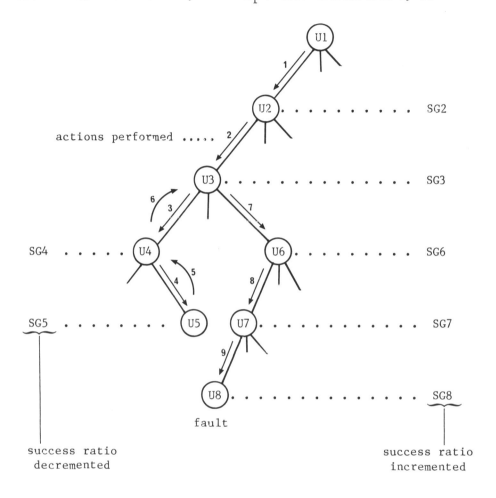

Fig. 10.9: Trace of a successful investigation.

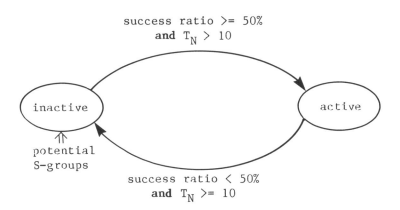

Fig. 10.10: S-groups status transition diagram.

Furthermore, for every successful investigation, each drop in level (en route to the fault) that was achieved by acquiring additional symptom information from the engineer is used to identify potential S-groups. In such cases the intersection between the involved observed symptoms set and K-group of symptoms forms a potential S-group for the relevant subunit, since had that S-group been in existence, the system would have dropped to the next lower level without having to suggest actions to the engineer. Referring to figure 10.9 the intersection between the K-group of unit U3 and the symptoms having been observed by the end of step 1 of the investigation forms a potential S-group for U3. If a potential S-group already exists but in the inactive status (i.e. it could not be used by the diagnostic process to yield matches with the CS set) then it would have its N_S and N_T values incremented by 1; otherwise it would be inserted in the space of inactive S-groups with both its N_S and N_T values initialised to 1.

Finally S-groups are demoted and promoted as depicted in figure 10.10.

REFERENCES

Addis T.R. (**1980**): "Towards an 'expert' diagnostic system", **ICL Technical Journal**, pp. 79-105.

Addis T.R. and **Hartley R.T.** (**1979**): "A fault finding aid using a content addressable file store", ICL Research and Advanced Development Centre, **tech. report 79/3,** Stevenage, U.K.

George F.H. (**1977**): "The CRIB handbook: an introductory description of CRIB". Unpublished.

Hartley R.T. (**1981**): "How expert should an expert system be?", Proc. **IJCAI—81,** pp. 862—867.

Hartley R.T. (**1984**): "CRIB: computer fault-finding through knowledge engineering", **IEE Computer,** pp. 76—83.